T0345147

TAKING THE LEAD IN PATIENT SAFETY

TAKING THE LEAD IN PATIENT SAFETY

HOW HEALTHCARE LEADERS INFLUENCE BEHAVIOR AND CREATE CULTURE

THOMAS R. KRAUSE, Ph.D. AND JOHN H. HIDLEY, M.D.

FOREWORD BY DIANE C. PINAKIEWICZ, M.B.A.
PRESIDENT, NATIONAL PATIENT SAFETY FOUNDATION

WILEY

A JOHN WILEY & SONS, INC., PUBLICATION

Published by John Wiley & Sons, Inc., Hoboken, New Jersey.
Published simultaneously in Canada.

Library of Congress Cataloging-in-Publication Data:

Krause, Thomas R., 1944-
Taking the lead in patient safety : how healthcare leaders influence behavior and create culture / Thomas R. Krause, John Hidley.
p. ; cm.
Includes bibliographical references and index.
ISBN 978-0-470-22539-4 (cloth)
1. Medical errors—Prevention. 2. Health services administrators. 3. Patients—Safety measures. I. Hidley, John H. II. Title.
[DNLM: 1. Health Facilities. 2. Safety Management. 3. Leadership. 4. Medical Errors—prevention & control. 5. Organizational Culture. 6. Patient Care. WX 185 K91t 2009]
R729.8K733 2009
362.1068—dc22 2008035475

Printed in the United States of America.

10 9 8 7 6 5 4 3 2 1

Book Design & Layout: theBookDesigners | bookdesigners.com

Cover image © iStockphoto.com/Alex Nikada

Contents

Foreword by Diane C. Pinakiewicz, M.B.A. . . . . *ix*
Acknowledgments . . . . *xi*

Introduction 1
Think leadership 4
Think systems 5
Think strategy 6
Think culture 6
Think behavior 7
About this book 8

1 What Determines Patient Safety? 9
Why make safety happen? 14
What stands in the way of improved healthcare safety? 16
Whose job is it to take the lead? 29

2 Blueprint for Healthcare Safety Excellence 31
The working interface:
Where exposure to hazard can occur 35
Healthcare safety-enabling elements 44
Organizational sustaining systems 45
Organizational culture 48
The charge of the safety leader 51

3 Nine Dimensions of Organizational Culture 61
Measuring culture with the Organizational
Culture Diagnostic Instrument 63
Organizational dimensions: The four pillars of culture 72
Team dimensions 78
Safety-specific dimensions 80
Why do some organizations change
more readily than others? 85

4 Qualities of a Great Safety Leader 89
The Safety Leadership Model 92
Measuring leadership with
the Leadership Diagnostic Instrument (LDI) 95
Personal safety ethic 98
Leadership style 112

5 Leadership Best Practices 119
Vision 124
Credibility 126
Action orientation 127
Collaboration 128
Communication 129
Recognition and feedback 130
Accountability 131
Measuring leadership best practices with the LDI 133

**6 Changing Behavior
with Applied Behavior Analysis 135**
What is behavior change? 137
Antecedents, behaviors, and consequences 139
ABC analysis 141
Putting the tools to work in your organization 146

**7 Protecting Your Decision Making
from Cognitive Bias 149**
Tragedy on Mount Everest 151
Cognitive bias and healthcare safety 153
Biases of data selection 155
Biases of data use 162
Case study: Cognitive bias in manufacturing 169
Putting your knowledge to work 172

8 Designing Your Safety Improvement Intervention 177
The Leading with Safety process 180
Phase I: The Patient Safety Academy 182
Step 1: Gain leadership alignment on patient safety as a strategic priority 183
Step 2: Develop a patient safety vision 191
Step 3: Perform a current state analysis 200
Step 4: Develop a high-level intervention plan for phase II . . . 204

9 Launching Culture Change for Patient and Employee Safety 209
Phase II: Achieving safety throughout the organization 210
Step 5: Engage the organization in the Leading with Safety process 212
Step 6: Realign systems, both enabling and sustaining 216
Step 7: Establish a system for behavior observation, feedback, and problem solving 222
Step 8: Sustain the Leading with Safety process for continual improvement 224
Case history: Exemplar HealthNet 225
Leadership Coaching 234

10 NASA After Columbia: Lessons for Healthcare 239
NASA's approach to culture and climate transformation 243
Assessing NASA's existing culture and climate 245
BST's NASA intervention 250
Results at NASA 257
Lessons for healthcare 265

Bibliography 271
Index 279

FOREWORD

In the fall of 1996, an amazing thing happened at the Annenberg Science Center in Palm Springs, California. Key stakeholders from the healthcare community convened a meeting to consider the emerging issue of medical error and invited experts in safety and human factors from other high-risk industries to join in the deliberations. This signaled the beginning of a new era in quality improvement efforts for healthcare as we began to look at the issue from a new perspective and avail ourselves of expertise that heretofore had not been considered in our work. From this meeting the National Patient Safety Foundation was formed and began the process of defining the issue and embracing the expertise of outside industries and established science in doing so. The field of patient safety began to take definition.

From the early days of Annenberg, there has been keen recognition of the critical roles that culture and leadership play in patient safety improvement efforts. The initial calls for action issued by NPSF and others encouraged the move away from a "culture of blame" and emphasized the necessity for strong leadership on this issue from the C-suite. Over the past decade, these calls for action have continued to be refined as we understand more about the issues at play and the levers for change and improvement at our disposal. We now understand the importance of a just culture, high reliability, transparency, systems thinking, teamwork, change management, front-line leverage, patient and family engagement, and leadership from the bedside to the boardroom. Through all of this work, our industry continues to draw on the expertise of those who have developed successful approaches to safety improvements in other industries and those who have translated those successes into relevant and valuable applications for the healthcare field—companies such as Behavioral Science Technology.

In this book the founders of BST bring their 25 years of expertise in safety science to bear on the critical issue of culture change and the associated leadership imperative required to move the patient safety work forward. While the importance of culture as context to the work at hand is well understood in our field, it often presents itself as an intangible and challenges those who seek to improve it with an inability to understand and measure the success of their efforts. This book helps to demystify this work by providing a structure with which to assess and measure aspects of culture and the impact that leadership can have on it.

In describing the working interface, the authors make a distinction between culture and climate and define nine critical dimensions of organizational culture along with four cultural pillars and show the predictive relationships between culture and safety outcomes. Their Safety Leadership Model, Leadership Diagnostic Instrument and definition of safety leadership best practices are predicated upon evidence from their work in outside industries and provide tangible ways of assessing and focusing leadership approaches to driving the culture change necessary for safety improvements. The work provides a way to define components of culture in a manner that allows them to be more readily understood and, therefore, addressed. Practical descriptions of the relevance of applied behavioral analysis to leadership challenges are presented along with excellent examples of how cognitive bias plays into leadership decision making. The important connection between patient safety and employee safety efforts is made and taken into consideration in their comprehensive look at the entirety of the culture within which our efforts must take place.

At every turn, this book makes tangible the intangible and helps to provide points of connection from which to advance our leadership and culture improvement initiatives. The evidence-based models and applications are unique and an example of the importance of our expanded view of the toolbox available to our industry as it goes about this important work. It is a book to be embraced by those of us who embrace the work of patient safety and for students of leadership and culture across the field.

DIANE C. PINAKIEWICZ, M.B.A.
PRESIDENT
NATIONAL PATIENT SAFETY FOUNDATION

ACKNOWLEDGMENTS

This book is the result of an ongoing collaboration across a network of people: some within BST as well as clients and consultants to BST. We would like to thank some of those who have made this effort possible.

Consultants are always indebted to their clients. This work was made possible by the individual organizational leaders who saw that safety leadership and organizational culture in healthcare could and should be improved, and put resources on the line to do it. We have had the privilege of working with many great senior safety leaders in outstanding organizations, within healthcare as well as outside it.

There are too many to name them all here, but we would like to mention several leaders who stand out in the healthcare industry: Phil Newbold, president/CEO at Memorial Hospital and Health System in South Bend, Indiana, Ken Anderson, chief quality officer at Evanston Northwestern Healthcare, Peter Angood, VP/chief patient safety officer at the Joint Commission, Diane Pinakiewicz, president of the National Patient Safety Foundation, Linda Groah, executive director/CEO of the Association of periOperative Registered Nurses, Mike Silver, VP of strategy and development at HealthInsight, Marguerite Callaway, president of the Callaway Group, and Dave Hofmann, Hugh L. McColl Scholar in Leadership at the University of North Carolina at Chapel Hill.

We would also like to acknowledge the many physicians and other healthcare professionals who have shared their experiences and perceptions with us and without whose input the relevance of the book would be substantially diminished. We are especially indebted to Robert Cowan, cofounder and medical director at the Keeler Center for the Study of Headache, and Jeff Luttrull, a retinal surgeon in private practice.

The things we have learned from leaders outside healthcare have also been useful to the development of this book. We would like to thank Sean O'Keefe, who was NASA's Administrator at the time of our work with NASA in response to the Columbia tragedy, Paul O'Neill, the former chairman/CEO of Alcoa, Jim Dietz, executive vice president/COO at PotashCorp, Paul Anderson, chairman of the board at Duke Energy, and Tom Weekley, previously co-executive director of the UAW-GM Center for Human Resources.

We would also like to thank our colleagues at BST, some of whom contributed directly to the writing of the book and others indirectly, including Richard Russell, managing director, and Kate Cameron, director, at risk-e, our Australian affiliate. Portions of chapter 3, on the nine dimensions of organizational culture, were originally written in collaboration with Kim Sloat. The chapter on cognitive bias was adapted from material originally written by Rebecca Timmins. The leadership coaching sidebar in chapter 9 was drawn from a section written by Jim Huggett. The account of BST's work with NASA in chapter 10 is based on material written by BST President Scott Stricoff.

John Balkcom, a senior advisor and member of our Advisory Board, helped with the research underlying the book and also with the writing. Sharon Dunn, our healthcare practice leader, gave advice and counsel on how to make the book most relevant to healthcare leaders. Maija Rothenberg gave us the substantial benefit of her editorial expertise. Rebecca Nigel did the painstaking job of editing that pulled it all together, while Julie Knudsen and Marty Mellein designed the graphics and The Book Designers did the layout.

Our heartfelt thanks to everyone who contributed to the book's success.

INTRODUCTION

After the space shuttle *Columbia* disintegrated on re-entry over Texas in 2003, killing all seven crew members, the Columbia Accident Investigation Board examined the accident in "organizational" terms—i.e., how did NASA function organizationally with respect to safety outcomes? The investigation moved in this direction because of an early finding: NASA had known about the properties of the foam used to insulate the external fuel tanks of the shuttle. It had known the foam had risks that would prohibit flight. Nevertheless, the *Columbia* was launched—27 times before the event that ended in an accident. The board concluded that NASA had a "broken safety culture" and attributed the cause of the accident to "organizational factors as much as any technical failure."[1]

Patient safety errors are rising nationally[2] and merit urgent attention. But the numbers themselves do not tell healthcare leaders what they should *do*. What does it take to create a "safety culture"?[3]

We have been working on organizational safety—the broad relationship between an organization's culture and leadership and its safety outcomes—since 1979. We were asked to assess NASA's culture using the tools we have developed over the years and to recommend an intervention strategy that would address the issues the board had found. We quickly discovered a reluctance

---

1 Columbia Accident Investigation Board, *Columbia Accident Investigation Board Report* (August 2003), vol. 1, p. 9.

2 See, for example, "U.S. Hospital Errors Continue to Rise," *Washington Post*, April 2, 2007. Viewed at www.washingtonpost.com.

3 The phrase is a bit of a misnomer; see *Culture versus climate*, chapter 2.

on NASA's part to look outside its borders for insights and methodologies to address its issues.

Healthcare has shown a similar reluctance to learn from other industries. Many leaders and clinicians say healthcare is unique and must find its solutions from the inside out. In one sense this claim is valid: healthcare is indeed unique. In fact, "healthcare organizations" are not organizations at all, at least in the way that industrial organizations and government agencies are. A healthcare organization is more like a loose confederation of constituencies with a common objective. Even in the academic medical center, the affiliations and loyalties often attach more strongly to profession, discipline, and specialty than to the institution. This difference has major implications for creating the culture of safety that healthcare institutions know they must develop.

"Healthcare organizations" are not organizations at all; they are loose confederations of constituencies with a common objective.

None of this says that healthcare can't learn from other organizations and industries. It's already happening. Many healthcare leaders today recognize that their objectives can be accomplished by asking, "What is out there that we can learn from?'" They also want to know, "Is the work any good? Would any of it impress us?" The unspoken concern is often this: "The last thing we want to waste our time on is learning how to put up safety posters and give away prizes for coming to safety meetings. And isn't that the kind of thing safety in industry is all about?"

It's a reasonable question. Industry does have its share of slide-show-and-poster safety. But that is all a distraction. Industrial organizations have actually done incredibly good work creating a positive safety climate and an organizational culture that supports safety outcomes. Alcoa and ExxonMobil, for example, are leaders in their respective industries with regard to workplace safety. Employees in these organizations do dangerous work. Every day, ExxonMobil workers produce, process, and transport several million barrels of petrochemical products.[4] Alcoa employees work with molten aluminum oxide, 1,750-degree chemical baths, and 32,000-pound aluminum ingots.[5] At ExxonMobil in 2006 there were three workplace fatalities among its 82,000 worldwide employees.[6] At Alcoa there were two employee fatalities in 2006 among its worldwide workforce of 129,000.[7] The goal of both organizations is zero incidents.

Keeping industrial employees safe requires putting into place reliable systems that are operating well and used consistently across the organization. Employees must communicate and collaborate with each other, across departments, between shifts—even when their immediate interests may be in conflict. Keeping employees safe must be a value held in common by the culture of the organization.

The same is true of healthcare. If medical mishaps such as wrong-site surgeries, medication errors, patient falls, nosocomial infections, and employee injuries are to be reduced and eliminated, nurses have to be able to speak clearly and frankly to doctors, medical specialists must collaborate and communicate with treatment team members, and treatment team members must cooperate with supporting departments such as laboratory, nutrition services, pharmacy, and maintenance. Systems must be present and well employed to ensure the coordination of activities to provide safe outcomes. These are the same structural issues that leading industrial organizations and some government agencies have taken on successfully. Some companies have been working on these issues, and making substantial progress, for 20 years.

A physician friend who had the opportunity to see what leading industrial organizations now do in organizational safety shook his head and concluded, "Healthcare is in the Stone Age when it comes to safety." What he meant was, other organizations

---

4 ExxonMobil, "Managing Risk in a Changing Environment," June 1, 2007. Viewed at www.exxonmobil.com.

5 Andrew Eder, "For Workers at Alcoa, It's Always Safety First," *Knoxville News and Sentinel*, Sunday, December 9, 2007. Viewed at www.knoxnews.com. *Also see* www.alcoa.com.

6 *ExxonMobil Corporate Citizenship Report 2006*. Viewed at www.exxonmobil.co.uk.

7 Eder, 2007.

have solved problems that healthcare is just beginning to address. Accurate or not, the statement suggests (and we agree) that attention to patient and employee safety returns big rewards if done correctly. This book aims to shed some light on what that attention might look like.

A physician friend said, "Healthcare is in the Stone Age when it comes to safety."

Organizational change is difficult under any circumstances. Habits have been formed, subcultures have become influential, suboptimal ways of doing things have become the norm, and leadership is often out of touch, if not misaligned, with the day-to-day realities of operations. Changing a confederation of constituencies is even more difficult. Whereas an organization can address the needed change with a single voice, the coalition of professionals in a healthcare organization has many voices, each representing different perspectives and roles within the confederation. As a result, it is easy to see how independent one-off safety interventions at the operations level, though entirely well conceived, risk low rates of sustainability. More broadly, it explains at least in part why healthcare has found addressing patient safety issues so formidable.

We have been following patient safety since the Institute of Medicine (IOM) report in 1999[8] and employee safety in healthcare for many years before that. Our experience suggest five ways of thinking about patient and employee safety.

Think leadership

The optimal way to approach safety improvement in healthcare is to start by creating alignment among the leadership of the major constituencies: the board of directors, physician leaders, and the healthcare system leaders, including the CEO and his

---

8 Institute of Medicine, Committee on Quality of Health Care in America, *To Err Is Human: Building a Safer Health System*, eds. Linda T. Kohn, Janet M. Corrigan, and Molla S. Donaldson (Washington, DC: National Academy Press, 1999). Viewed at www.nap.edu.

or her direct reports. Can such a coalition be formed? We have seen numerous examples to suggest that it can—when the issue is patient safety.

All constituencies win when patient safety improves. As patient safety and employee safety improve, employee satisfaction improves, organizational citizenship improves, patient satisfaction improves, quality of care improves, malpractice costs decrease, and the overall reputation and financial security of the institution improve. These are documented research findings, not speculation.[9] In reality, ensuring the safety of patients and employees is the *only* way to keep the entire institutional organism healthy.

So the issue becomes one of leadership. A small group of leaders need to align themselves around this objective and create a tight, effective coalition among constituencies.

Think systems

Patient and employee safety is more about systems performance than individual performance. Adverse events result from complex processes embedded in the organization, not just from the actions of the individual clinician or technician. Patient safety

---

9 Thomas Krause, *Leading with Safety* (Hoboken, NJ: John Wiley & Sons, 2005).

Also see:

Kerry D. Carson, Paula P. Carson, Ram Yallapragada, and C. William Roe, "Teamwork or Interdepartmental Cooperation: Which Is More Important in the Health Care Setting?" *Health Care Manager,* 19 (2001): pp. 39–46.

A. Cohen and Y. Kol, "Professionalism and Organizational Citizenship Behavior: An Empirical Examination among Israeli Nurses," *Journal of Managerial Psychology,* 19 (2004): pp. 386–405.

Jacqueline A-M. Coyle-Shapiro, Ian Kessler, and John Purcell, "Reciprocity or 'It's My Job': Exploring Organizationally Directed Citizenship Behavior in a National Health Service Setting," *Journal of Management Studies,* 41 (January 2004): pp. 85–106.

Gerald R. Ferris, "Role of Leadership in the Employee Withdrawal Process: A Constructive Replication," *Journal of Applied Psychology,* 70 (1985): pp. 777–781.

Barbara Mark and David A. Hofmann, "An Investigation of the Relationship Between Safety Climate and Medication Errors as Well as Other Nurse and Patient Outcomes," *Personnel Psychology,* 59 (2006): pp. 847–869.

Tal Katz-Navon, Eitan Naveh, and Zvi Stern, "Safety Climate in Healthcare Organizations: A Multidimensional Approach," *Academy of Management Journal,* 48 (2005): pp. 1075–1089.

Mary A. Konovsky and S. Douglas Pugh, "Citizenship Behavior and Social Exchange," Academy of Management Journal, 37 (June 1994): pp. 656–669.

and employee safety are inseparable; both are produced by the same attributes of the organization's culture.

Root cause analyses of incidents show clearly that while an individual can be blamed (e.g., the nurse who overrode the medication dispensing machine), the real cause of any incident is almost always a failure of systems. Yes, if the nurse would have called the right person and asked the right question, the incident might have been avoided. But if the system doesn't provide for adequate staffing, if overrides are commonplace and seen daily by treatment team supervision and any leader who bothers to look, if training is inadequate, if relations among departments and individuals are tense and strained, then it is profoundly unjust to look backward after an incident that was predictable, given the systems factors, and blame the employee.

Root cause analyses find a common pathway in virtually every adverse event that ties specific behaviors to its cause and therefore to its prevention. These behaviors, however, must not be understood in isolation. Rather, they occur in the context of a convergence of multiple systems and cultural influences. The responsibility of providing adequate systems belongs to the leadership of the institution.

Think strategy

Fixing patient safety requires a strategy, not a set of one-off tactics. Too often safety comes after efficiency, after economy, and after profit. Too often in professional group meetings or in board meetings, safety is neither the first thing on the agenda nor on the agenda at all. Displaying safety as one of the 65 measures on an institutional dashboard is no better: safety keeps its low prominence instead of commanding its rightful central position as a strategic value to the organizational leadership at all levels.

Think culture

Leaders create culture with their every thought, word, and deed. In fact leadership predicts culture—and culture predicts safety outcomes. The data demonstrate these relationships (see chapters 3 and 4). Since leadership shapes culture, and culture predictably defines the likelihood of exposure to harm, leaders are obligated to take action consciously and continually to mitigate hazard. With this knowledge comes the responsibility to act on what they know. The fundamental ethical error related to patient safety occurs when leaders know how to minimize exposures to harm but don't take action to make it happen. Healthcare leaders can know, sometimes do know, and always ought to know how climate, culture, and systems shape the behaviors manifested in their organizations.

So, if leadership creates culture, and if you are a leader who thinks the culture needs to change... what is implied?

Think behavior

Learn to think about patient safety in behavioral terms. Not everyone else's behavior—your own. Becoming an effective safety leader means finding the specific relationship between your actions as a leader and the state of patient safety in your institution, both organization-wide and in your own functional area of responsibility. Once you understand that relationship, you know how to change the behavior—to everyone's benefit.

> Becoming an effective safety leader means finding the specific relationship between your actions as a leader and the state of patient safety in your institution.

Before we can figure out how to get the nurse to stop overriding the dispensing machine, we first need to explore the likely gap between intentions and results, not with blame or fault-finding but by measuring behavior. There are well-tested methods for doing this sort of empirical observation. We use the Leadership Diagnostic Instrument (see chapter 4), which provides leaders with 360-degree feedback on their leadership style and best practices. Then we work with the leaders on developing specific behavioral strategies to increase those aspects of leadership known to correlate to good safety outcomes.

About this book

In *Leading with Safety* we reflected on our work with industrial and governmental organizations. Healthcare leaders then encouraged us to summarize the findings of our work in a way that would support leaders who are serious about improving patient and employee safety. We were pleased to take on the writing of this book, which covers the same principles as *Leading with Safety* but has been updated to reflect current research and our most recent insights.

Healthcare organizations are pliant—their qualities can be changed.

We approach the issue of patient safety not as yet another critic eager to take potshots at hard-working professionals and organizations but as sympathetic colleagues. We know the challenges and face them ourselves in the organization we run as well as those with whom we consult. We know that positive change can happen. We've seen it time and time again. The good news is that healthcare organizations are pliant and actionable. Their qualities can be measured and compared with a normative database of comparably complex organizations in both healthcare and other industries—and their qualities can be changed.

Once you know what's possible, we believe you will see your obligation to act. We share a common goal: the health and well-being of the millions of patients who put their lives in our collective hands.

CHAPTER ONE WHAT DETERMINES *PATIENT SAFETY?*

WHAT DETERMINES PATIENT SAFETY?

This book is for leaders who believe that patient safety is the right thing to do and who want to know how to make a difference in their healthcare organizations. It is for the people who influence not just the safety of patients but also that of team members providing patient care—imperatives that go hand in hand.

There is reason to be optimistic about one's ability to improve both. This book provides a roadmap to help you, as a leader, improve the level of safety in your organization. In this first chapter we explore what motivates great safety leaders and identify some of the sources of resistance you're likely to encounter.

Understanding the complexity of healthcare safety and the potential resistance to it will forewarn and forearm you and help you prioritize your efforts. Your specific role as a leader in healthcare safety is to discern the safety issues, define the terms of your organization's engagement with them, and use the issues to mobilize the sometimes competing and conflicting constituencies that must cooperate to create a culture that supports safety. Your guiding principle for executing this role is this: All patients have the right to expect not to be injured by the healthcare delivery system. To do no harm[1] is our first duty.

---

1 "As to diseases[,] make a habit of two things—to help, or at least, to do no harm." Hippocratic Corpus, *Epidemics,* book I, chapter II. From *The Yale Book of Quotations,* ed., Fred R. Shapiro (Yale University Press, New Haven and London, 2006). This injunction, which likely came down to us through Galen, has been espoused in the modern age by prominent medical practitioners such as Florence Nightingale.

We begin with a case history unlike those in most other texts about medical errors. In the case of CH, a person of our acquaintance, everything went well—in large part because she felt empowered to speak up at the right moment.

> CH was a 28-year-old, 2-weeks pregnant, white female mother of two who visited her physician and was seen by his nurse practitioner on a Friday in September 1999. She complained of flu-like symptoms and a lump in her neck. Her physical examination was within normal limits except for a swollen, mobile, nontender right cervical lymph node and a rash on her elbows. Blood was drawn and she was sent home. The following Tuesday, the doctor told her that her results did not look quite right and requested that she return for repeat blood work. The repeat test confirmed that she had a very high white count, and the doctor diagnosed leukemia. He sent her to an oncologist, who detected the Philadelphia chromosome and diagnosed chronic myelogenous leukemia.
>
> The patient had a spontaneous miscarriage the Friday after she was first seen and was started on hydroxyurea shortly thereafter. A bone marrow transplant was scheduled for mid-January 2000. However, in November 1999 the patient was found to be pregnant again and elected to carry the baby to term. Since she was unable to take leukemia medications during the first trimester, a right atrial catheter was inserted so that, if necessary, excess white cells could be removed from her blood. After the first trimester the atrial catheter was removed, hydroxyurea was restarted, and she progressed through an uneventful term pregnancy with the birth of a normal, healthy baby.
>
> On January 2, 2001, the patient entered the City of Hope Hospital in Duarte, California, for a bone marrow transplant. There she underwent eight days of treatment with a chemotherapy cocktail to kill her bone marrow. She was transfused with donated marrow on January 11, started on prednisone, and observed for graft versus host disease. She was discharged on February 28 and instructed to live within 15 minutes of the hospital.
>
> Although she had several bouts of fever that required rehospitalization, she never developed graft versus host disease.

> Her prednisone was discontinued. She is now checked annually, and as of January 2007 she was leukemia-free for 6 years.

So far, the story of CH's medical care is both miraculous and ordinary. Despite the many danger points in her treatment, mistakes have been avoided and the outcome is happy. The case also reveals the miraculous power of modern medicine: the ability to examine human chromosomes; to place a tube into the heart of a

> It is important to maintain a realistic perspective, especially as a patient safety leader. The odds of patients being helped far outweigh their chances of being harmed.

living person; to withdraw her blood, remove excess white cells from it, and return it to her body; to kill a patient's bone marrow and replace it with someone else's, thereby giving her a new immune system; and to cure leukemia. The story is ordinary in that interventions like this happen every day.

Today, while the American healthcare system receives constant criticism, it is important to maintain a realistic perspective, especially as a patient safety leader. With respect to any given patient, there are three strong likelihoods:

- When patients go to their doctors today, their chances of being helped are high—in fact, their chances are better now than at any previous time in human history.
- The chance remains small that a patient will be injured by a physician or by the healthcare system.
- Thus, the odds of patients being helped far exceed their chances of being harmed.

It is important to be clear about these probabilities up front. Patients should not be afraid to go to their doctors or to be treated in a hospital. Healthcare leaders and workers take very seriously their responsibility to help, and the idea of doing harm is abhorrent to them. On the other hand, patient safety is not a trivial concern. Despite the best of intentions, sometimes patients are indeed injured by those who are trying to help them and by vulnerabilities in the healthcare delivery system itself.

This risk was dramatized and thrust into the public's awareness by the 1999 IOM report, *To Err Is Human: Building a Safer Health System,*[2] which states that in any year, 44,000 to 98,000 Americans die because of medical errors. As Robert M. Wachter and Kaveh Shojania explain in their excellent book, *Internal Bleeding: The Truth Behind America's Terrifying Epidemic of Medical Mistakes,* more people die each year in the United States as the result of medical errors than from AIDS and breast cancer combined.[3] Others have pointed out that this terrible outcome is the equivalent number of lives that would be lost if a Boeing 767 full of passengers crashed every day of the year.

Although these numbers are controversial, the situation clearly warrants action. There is ample room for improvement in patient safety. Medical leaders—among them, physicians, nurses, pharmacists, administrators, regulators, politicians, third-party payers, and concerned organizations and agencies such as the National Patient Safety Foundation, Joint Commission, Agency for Healthcare Research and Quality, and others—have taken the IOM report very seriously and are searching for ways to improve healthcare safety. Everyone is in agreement that:

- Even one unnecessary death within the healthcare delivery system is unacceptable. Yet many such deaths occur each year.
- The problem is urgent; we cannot turn a blind eye to it while conducting business as usual.
- Leadership is needed to change the culture, behaviors, and processes that allow medical errors to happen.

---

2 Institute of Medicine, Committee on Quality of Health Care in America, *To Err Is Human: Building a Safer Health System*, eds. Linda T. Kohn, Janet M. Corrigan, and Molla S. Donaldson (Washington, DC: National Academy Press, 1999). Viewed at www.nap.edu.

3 Robert M. Wachter and Kaveh Shojania, *Internal Bleeding: The Truth Behind America's Terrifying Epidemic of Medical Mistakes* (New York: Rugged Land, 2005). This book, a must-read for healthcare safety leaders, offers ample facts and ideas and provides an excellent background on the issues involved.

But as we shall see, the challenges faced by any healthcare safety leader are substantial, because the system that delivers healthcare is so highly complex. Moreover, it is more often the human element—leadership, culture, and behavior—rather than the science that proves to be the weak link in the chain of healthcare delivery and patient safety.

Why make safety happen?

Over the past 20 years, the authors and their associates at BST have helped more than 2,000 organizations in 40 countries with many different cultures and languages build strong organizational cultures that support safety and significantly improve

It is more often the human element—leadership, culture, and behavior—rather than the science that proves to be the weak link in the chain of healthcare delivery and patient safety.

their safety performance. We find that what motivates healthcare leaders differs little from what motivates the safety leaders in non-healthcare companies.

Ethical considerations

The primary motivation of safety leaders everywhere is ethics. Business leaders don't want to injure their customers or employees; likewise, healthcare leaders don't want to injure their patients or their employees. Fulfilling the moral obligation to ensure a safe setting for both employees and patients builds employee loyalty, reinforces the healthcare organization's reputation, and legitimizes its claim of serving the local community—thereby authenticating its right to a license to operate. Like their industrial counterparts, healthcare leaders also have business reasons for pursuing patient safety that are less altruistic but no less valid: to address business considerations, to align the organizational culture, because they recognize that safety is good strategy, or all the above.

Business considerations

Modern healthcare is big business. The bottom line counts. The cost of uncompensated services due to adverse events comes right out of the organization's reserves. Although an adverse event may on occasion increase revenue or in some instances decrease short-term costs, it is more likely to increase the organization's legal exposure and damage its reputation in the community, thereby diminishing its ability to compete.

On the upside, a positive bottom line—whether earned by a for-profit owner or by a nonprofit corporation in the public interest—enhances the long-term viability and sustainability of a healthcare organization. The senior leadership, whether clinical or administrative, bears chief responsibility for the sustained capacity of the organization both to deliver care and to ensure the safety of patients and staff. Although financial viability may not come first, more than one nonprofit leader has been heard to say, "If no margin, then no mission."

Aligning organizational culture

In any industry, healthcare included, directly addressing safety indirectly helps with many other organizational problems. Many healthcare organizations are fragmented along departmental lines; many, if not most, suffer vested interests and fiefdoms. They also tend to suffer from rivalries and conflicting professional agendas. Patient safety is one area in which all parties have a common interest, an area where departments can learn to work together to create a common culture and a strong safety climate.

From working with one organization after another, we now know that both culture and climate can be defined and measured concretely. We also know that although culture (the way we do things around here, rigorously observed) and climate (what gets noticed and recognized most immediately and prominently) are both important to safety, it is leadership that creates them, and leadership that distinguishes those organizations exhibiting good safety performance from those that excel.

We have also learned that aligning leadership constituencies can pay off in other ways. Alignment consists of getting all the constituencies to see the issue from the same perspective and to think about it in the same ways, thereby creating a common language and a shared way of understanding safety issues. This commonality makes for less conflict, easier and more successful communication, and smoother operations. These factors all facilitate both safety and productivity.

Finally, building a strong safety climate makes meeting healthcare safety regulations much easier. We place regulatory compliance last because we regard it as necessary but woefully insufficient to the avoidance of hazards for patients and employees.

Safety is good strategy

Recognizing the many advantages of successfully attending to this issue, some forward-looking healthcare organizations already regard healthcare safety and the idea of being recognized as a world-class healthcare provider as a strategic issue: it pays off in reputation, referrals, and reimbursement schedules, while enhancing the organization's ability to attract and keep the best talent in healthcare delivery and leadership.

While these business advantages are strong, the most common and fundamental motive of great safety leaders is that they see clearly that improving safety is the right thing to do. They hold fast to the principle that we have an obligation to ensure that first, we do no harm.

What stands in the way of improved healthcare safety?

The complexity of modern medicine is both miraculous and a source of many of its errors. This complexity makes it difficult for a leader to have a clear grasp of where to begin. It also engenders a resistance to change that those aspiring to higher levels of healthcare safety may encounter as they work to improve the safety of their organizations. You may encounter these resistances not only in others but in yourself. The patient safety problem can seem overwhelming, that is just too daunting, so perhaps you should just focus on issues that will yield more easily to your efforts.

This conclusion would be mistaken. There is great variability among organizations in healthcare safety performance. For example, when HealthGrades, Inc., analyzed publicly reported information on Medicare patients in all U.S. hospitals annually during 2003 through 2005, it found the following:[4]

- There were wide, highly significant gaps in individual PSIs [Patient Safety Indicators] and overall performance between the Distinguished Hospitals for Patient Safety and the bottom-ranked hospitals.
- Medicare patients in the Distinguished Hospitals for Patient Safety had, on average, approximately a 40% lower occurrence of experiencing one or more PSIs compared with patients at the bottom-ranked hospitals. [That is, these patients were 40% less likely to suffer the occurrence of one or more PSIs compared with patients in the bottom-ranked hospitals.] This finding was consistent across all 13 PSIs studied.
- If all hospitals performed at the level of Distinguished Hospitals for Patient Safety, approximately 206,286 patient safety incidents and 34,393 deaths of Medicare patients could have been avoided while saving the United States approximately $1.74 billion during the period 2003 to 2005.

So there are definitely ways to do things better. Furthermore, leaders who have tackled this issue have been able to show results rapidly, and they have found that even small improvements have substantial short-term payoffs. Table 1–1 summarizes some of the efforts described in the Commonwealth Fund's 2006 report, *Committed to Safety: Ten Case Studies on Reducing Harm to Patients.*[5]

---

4 HealthGrades Fourth Annual Patient Safety in American Hospitals Study (Golden, CO: Health Grades, Inc., 2007), pp. 2–3, 5, and appendix.

5 Adapted from Douglas McCarthy and David Blumenthal, *Committed to Safety: Ten Case Studies on Reducing Harm to Patients* (New York: Commonwealth Fund, April 2006). Viewed at www.commonwealthfund.org. Used with permission.

TABLE 1-1. TEN PATIENT SAFETY INTERVENTIONS AND RESULTS.

INTERVENTION	OBSERVED IMPROVEMENT
Accelerate patient safety in a 569-bed, level I trauma center through a multifaceted culture change program involving setting and monitoring behavioral expectations, enhancing analytic capabilities, and streamlining and focusing on critical policies.	• 42% increase in expected communications behaviors. • 50% reduction in events of harm per 10,000 adjusted patient days when culture change strategies were applied system-wide.
Lead organizational cultural change in Veterans Administration hospitals by empowering local facilities and front-line staff with proven tools, methods, and initiatives for patient safety improvement.	• 30-fold increase in internal safety incident reporting. • 100% increase in perceived preventability of safety events studied by root cause analysis teams.
Initiate a preoperative safety briefing and perinatal patient safety project as part of a program of organizational learning to promote effective teamwork and communication in high-risk areas in an integrated group-model health maintenance organization with 8.2 million people enrolled nationally.	• A near doubling in the proportion of operating room staff reporting a positive teamwork climate. • Two-thirds reduction in the turnover rate among operating room nursing staff.
In a 295-bed community hospital that annually treats 250 patients in its cardiac surgery program, use collaborative rounds involving all members of the care team with the patient and patient's family to proactively identify and prevent potential errors and safety threats.	• 56% lower than expected risk-adjusted mortality among cardiac surgery patients. • 15% to 32% higher staff ratings of teamwork and work satisfaction compared with traditional rounds.
In a 489-bed acute care hospital, establish rapid response team to intervene early with patients showing signs of medical deterioration before they suffer acute crises.	• 60% decrease in emergency calls for respiratory arrest. • 15% decline in cardiac arrests. • 3.95% reduction in hospital mortality rate.
In a 14-bed oncology surgical intensive care unit (ICU) and a 15-bed surgical ICU within a 900-bed academic medical center, implement a comprehensive unit-based safety program that empowers staff to identify and eliminate patient safety hazards following eight action steps.	• 49% to 91% increase in the proportion of ICU staff reporting a positive safety climate. • Elimination of 43 observed catheter-related bloodstream infections, saving eight lives. • One-day reduction in average ICU length of stay, saving an estimated $2 million annually.
In more than 40 ICUs in diverse community hospitals nationwide, focus all members of the care team on adhering to a "bundle" of evidence-based care practices associated with improved patient outcomes.	• 29% to 41% reduction in combined rates of ventilator-associated pneumonias. • 11% to 15% decrease in average lengths of stay across participating ICUs. • 18% lower mortality.

INTERVENTION	OBSERVED IMPROVEMENT
Specify best practices, eliminate variation from standards, and work toward ideal performance in a medical ICU and a cardiac care ICU in an 829-bed academic health center.	• 76% reduction rate of central-line-associated bloodstream infections, saving 18 lives per year. • $2 million savings by reducing unreimbursed costs of care.
For an alliance of more than 200 not-for-profit hospitals and health systems, develop a trigger tool to measure the incidence and kinds of adverse events, so as to prioritize areas for improvement, design appropriate interventions, and track the effect of changes over time.	• 50-fold increase in detection of adverse drug events as compared with other common methodologies such as incident reporting, pharmacy interventions, or billing codes.
Reduce adverse drug events by improving the process of medication reconciliation, the safe use of high-risk medications, and the reliability of medication dispensing in a 165-bed acute care hospital.	• 10-fold reduction in detected adverse drug events. • 8% improvement in perceived safety culture among hospital staff.

Source: Commonwealth Fund.

The interventions highlighted by the Commonwealth Fund report are sometimes isolated, sometimes funded demonstration projects, and often not integrated solutions that form an overarching organizational strategy. The use of one-off approaches, while informative, often threatens the achievement of sustainable results. Still, if the efforts are sustained, the financial benefits implied in Table 1–1 are clear: reductions in unreimbursed costs of care, improved patient progress, enhanced bed utilization, and more efficient and effective treatment team involvement.

So the situation is not hopeless. Far from it! But it is complex. To begin to understand this complexity, let's take a look at the safety dilemma as seen through the eyes of some of healthcare's key constituencies.

Safety roadblocks from the patients' point of view

Patients complain that they can't afford the escalating costs of healthcare or medical insurance. Many people lack medical insurance and must seek treatment in busy emergency rooms where they are triaged under pressure and may or may not receive timely treatment. But even for patients with insurance, the system does not always provide a smooth ride. Patients complain that when their previously approved provider is dropped from the insurance company's authorized provider list (because, for example, the provider's group was unable to negotiate a satisfactory contract), the patient must either pay more out of pocket or find a new provider.

Finding a new provider can be overwhelming, especially for the elderly and infirm. Self-employed patients and young patients can have a similarly difficult time. Patients complain that they have no reliable way to determine who is a good doctor. When diagnosed with a serious illness, they find it difficult to assess the quality of the hospital where they will be treated.

Even when their treatments are successful, patients sometimes experience mistakes in their medical care. Let us continue with CH's story:

> CH received frequent chest x-rays to make sure she was not developing pneumonia during the period in which she lacked a functioning immune system. On one such x-ray visit, she was not offered a protective apron for her abdomen until she spoke up and requested one. The radiology technician explained that he had thought she didn't need one because the radiation dose she was about to receive from a chest x-ray was slight compared with the much higher dose he (incorrectly) assumed she would soon be receiving in the radiation treatment for her leukemia. She told him that there was no such plan to give her radiation treatment.

This interaction between CH and the radiology tech was an example of not only a breakdown of communication between departments but also an incorrect supposition on the part of the technician.

Why did CH speak up? In part, it could be her personality. Perhaps CH is naturally assertive and does not hesitate to ask that her needs be met. It could have been the personality of the radiology tech, who communicated through his words and demeanor a willingness to respond. In telling us her story, CH said she spoke up because the culture of the City of Hope hospital where she was being treated had been so clearly and consistently on her side in her fight against leukemia. The people at City of Hope obviously wanted to do everything they could to ensure her survival and support her well-being. Without such a culture, this woman of childbearing age might have said nothing—and might have received unnecessary radiation to her reproductive organs.

A strong, healthy culture is an essential ingredient in healthcare safety. One indicator of a strong, healthy culture is that everyone—whatever their position in or relationship to the organization, whether patient, vendor, nurse, CEO, janitor, or visitor—feels empowered to step outside the usual reporting hierarchy to raise a safety concern or call a halt to an unsafe activity before an error occurs.

In addition to communication problems, relationship problems can also cause safety issues. Patients complain that doctors are impersonal, arrogant, or brusque; that doctors don't give patients enough time to unburden themselves of their concerns; that doctors don't really listen. Patients complain that their doctors say things the patients don't understand and give instructions that are impractical to follow or hard to remember. These communication and relationship issues create the climate for a malpractice perfect storm—a storm that can brew even in the absence of medical errors, because of an undesirable patient outcome.

Safety roadblocks from the physicians' point of view

Hospitals, physicians, and other clinicians deal in matters of disease, disability, and death. Many of the patients are frail; some are only a step away from the grave. At the next unfortunate turn of events, some will suffer even greater disability; others will not survive.

Are all adverse events preventable? The answer depends on the definition of "adverse event." Some patients will die in the natural course of events, irrespective of the quality of treatment provided. But this by itself does not mean that some number of medical errors are inevitable.

One of the criticisms of the IOM report is that it does not do a good job distinguishing between adverse events and deaths that were not preventable. This ambiguity is important because it can leave the treating staff feeling guilty, undermine their professional confidence, and provide a loophole for their conscience, leading to the unfortunate attitude that some number of untoward events is inevitable. The important issue is not blame but how to design systems and human factors so that they mitigate the hazard to the patient. Improving these factors is a continuous process.

This case, originally reported in *Hospital Pharmacy,*[6] was part of a course presented to pathologists by a medical malpractice insurance company.

> A patient underwent a surgical procedure to remove a cancerous eye. A nurse set everything up for the case. To prepare a container for the surgical specimen, she poured glutaraldehyde (preservative) into a medicine glass and placed it on the sterile field. Excess spinal fluid had been removed from the patient to reduce cerebral pressure because the malignancy had spread to his brain. The spinal fluid was in another unlabeled, identical, medicine glass also on the sterile field.
>
> Decreased spinal fluid can cause severe headaches. So near the end of the procedure, the anesthesiologist filled his syringe with what he thought was the patient's spinal fluid and injected it into the spinal canal. Unfortunately, the clear fluid he injected was not spinal fluid but glutaraldehyde. Glutaraldehyde is extremely toxic. The patient's nervous system was irreparably poisoned and he died.

A patient is dead. Whose fault is it? The aftermath among the team members from this tragic accident is unknown, but we can imagine the devastation and guilt. The anesthesiologist may have wondered why in the world the nurse didn't explain what was in each medicine glass; the nurse, paying attention to her own tasks, may have wondered why he didn't ask. Both were working in good faith, trying to do their jobs. One may have blamed the other, or so we speculate. The accident may have had career implications for those involved.

Whatever the ethical and psychological outcomes for the team, from a legal standpoint, under the "captain of the ship" doctrine,

---

6 *Hospital Pharmacy,* Medication Error Reports column, July 1989. *Note:* The details of this presented case were recalled from memory and may differ from actual events.

the anesthesiologist was considered to be in charge of everything that went on in the operating room.

The anesthesiologist may have concluded that despite medicine's marvelous technological advances, human errors like this one are not always preventable. From the point of view of the individual provider working in isolation, this perspective is understandable. From a systems and cultural perspective, it is a different matter altogether (more on this in chapters 2 and 9). While it is obviously our responsibility to prevent adverse events, it is also our duty to learn how to prevent adverse events that are not currently perceived to be preventable. The latter may be the harder challenge.

Adverse drug events[7] and mistakes that occur due to communication[8] problems and handoffs are among the most frequent types of medical errors. To the doctor these often appear to be inexcusable errors that have occurred because somebody else wasn't paying attention or failed to speak up. But there is always more than enough blame to go around, and a doctor may have blinders on about his or her role in the event. Perhaps it was her poor handwriting or his terse verbal style that created the occasion for error. When asked what she thought was the solution to such errors, one surgeon replied, "A new generation of docs. I'm not going to change. I'm the captain of the ship, not a team cheerleader, and I certainly don't want someone looking over my shoulder in the OR."

Physicians deplore mistakes and inefficient organizational systems that let them down and injure their patients. But few doctors are trained in systems thinking, culture development, or process control methods, and they often feel that these problems are beyond their control, i.e., the responsibility of other people. The attitude is, "My job is to practice good medicine. Nurses should accurately carry out my orders. Administrators should see to it that hospital systems function properly."

The doctor who doesn't think that his or her professional identity encompasses a leadership role in patient safety is a part of the problem. Whether or not a doctor acknowledges

---

7 Institute of Medicine, *To Err Is Human*, pp. 32–35. Viewed at www.nap.edu. See also Jane E. Brody, "To Protect Against Drug Errors, Ask Questions," *New York Times*, January 2, 2007, p. D7.

8 See The Joint Commission's "Root Causes of Sentinel Events" report at www.jointcommission.org.

it, every physician is a leader in setting the tone of the culture of the organizations within which he or she functions. For better or worse, the doctor's impact extends deeply into the organization. As clinicians, we should acknowledge our impact and join others in learning how to improve the organization and its culture.

Doctors often cite unreasonable production pressure as leading to medical errors, as well as the necessity to practice wasteful, defensive medicine in order to avoid malpractice exposure. Doctors often think that they don't have time to relate to their patients on anything but a technical level.

The explosive growth of medical knowledge further complicates the professional lives of clinicians and puts additional strains on their time. Some doctors complain that they don't even have time to do a thorough job of sustaining technical mastery. An internal medicine colleague recently confided that she struggles just to stay current on the cardiovascular literature and does not have time to read other journals. Much of her practice does consist of patients with cardiovascular disease, but what about those who have other problems?

Excessive demands on their time and an unjust malpractice system damage physician morale. Some doctors have even stopped practicing medicine over these issues; others refuse to perform certain procedures. There are communities, for example, where none of the obstetricians will deliver babies and no psychiatrist will perform electroconvulsive therapy. If you have a complicated pregnancy or suffer an intractable depression, good luck!

Another major consequence of defensive practice is maintaining the public appearance of perfection. One of the many possible repercussions of this façade is the failure to learn from errors, and certainly the failure to share learnings with others.

But, as often happens in healthcare, the situation is even more complex than it at first appears. A doctor's knowledge of the latest procedure or new medication does not always benefit the patient. Take, for example, a new, scientifically validated clinical best practice protocol. Should the physician adopt it? Not necessarily. It may contain procedures or call for the use of medications with which the physician has little familiarity and no experience. It is the patient who pays the price of the physician's learning and incurs the cost of mistakes while the doctor gains the experience necessary to use the new protocol proficiently. Sometimes it serves the patient far better for the doctor to stick with his familiar methods. Using them, the doctor can draw on an extensive body of experience to recognize when the patient is vulnerable to complications, failing to respond as expected, developing a significant side effect, or in need of a different approach or a referral.

This recitation is just to underscore that the situation is complex. It is not to say that best practice protocols have no place in clinical practice. We have heard protocols rejected by colleagues who thought they were defending good practice against "cookbook medicine." The issue is not cookbook medicine versus good medicine. Rather, it is the difficulty of the decisions facing physicians. Some decisions are easy because the right way to proceed is uncontroversial and clearly known: e.g., routinely examining the feet of patients with severe diabetes, using aspirin in cardiovascular disease, or administering tissue plasminogen activator to dissolve blood clots within three hours of the onset of ischemic stroke symptoms. These simple best practices should be a standard for all physicians.

The ease or difficulty of a decision is a function of how much is known about the specific patient and how far the relevant science has advanced. There are decisions for which many variables are involved and a lot is not known. These cases draw heavily on the clinician's experience, and in these cases the value of clinical protocols is less clear-cut.

Much criticism has been leveled against doctors for relying too heavily on their individual experience when they make critical medical decisions. Important studies of cognitive bias show the dangers inherent in this kind of decision making, and we will discuss these vulnerabilities at some length in chapter 7. Nevertheless, cognitive science has shown that experience-based decision making is what distinguishes experts from amateurs and adult decision processes from those of teenagers. Experience is also what distinguishes all human thought from computerized imitations, and it is why clinicians generally perform better than software algorithms. Computerized decision aids based on good science are undoubtedly part of a solution for decreasing medical errors, but we should not be too eager to replace the practitioner's experience with an algorithm.

The professional lives of physicians are complicated by regulatory intrusion and third-party payer interference with medical decision making, by the concomitant growing paperwork burden, and by the resultant impact on the physician's time, autonomy, and bottom line. A pediatric patient who had been physically abused was refused treatment by a third-party payer. When one of the authors appealed the decision on the child's behalf, the reviewer for the third-party payer, a retired nephrologist, told the author that in his opinion, the child just needed a good spanking! The patient was again refused authorization. Was this poor decision making on the part of the nephrologist a medical error? It was certainly a problem in the healthcare delivery system, one that further damaged the patient.

Safety roadblocks from the nurses' point of view

Nurses generally spend more time with patients than any other professionals in the healthcare system. Because of their proximity to patients they often possess detailed knowledge of medical errors and their causes. Again, because of frequent interaction with patients, nurses are also the last line of defense against medical errors and are in a position to catch errors before any damage is done. What do nurses say about medical errors and patient safety?

The first thing they usually mention is time pressure. Nurses complain that they have more patients than they can manage, that teamwork is poor, morale low, and burnout high. They feel their availability for bedside patient care is compromised by a growing body of administrative tasks.

Confusion is another issue. It can stem from any of several sources:

- Similar names and packaging for different drugs
- Storing different but similar-looking drugs in the same place
- Frequent changes in drug suppliers, which results in a drug's being packaged in a different container with every new supplier

- Equipment from different manufacturers that does the same job, but operates differently and has different readouts
- Different ways of doing things at different hospitals (many nurses are temps)
- Every doctor having his or her own preferred way of dealing with specific medical issues

When asked what stands in the way of improved patient safety, nurses often point to administrative and physician indifference or complacency, to arrogant and disruptive physician behavior, to physicians who are unwilling to make themselves available when needed, to a punitive response to the voicing of errors, and to the lack of cooperation between departments. They sit at the patients' sides more often than other practitioners and often find themselves at the eye of a confusing and daunting storm.

Safety roadblocks from the administrators' point of view

Administrators cope with increasing regulatory demands, exposure to unreasonable legal liability, and being squeezed in their contracts with large third-party payers. They often face inadequate resources, stiff competition from the hospital across town, and hard-won profits. They are concerned about the nursing shortage, poor morale, and high turnover, particularly among critical care nurses. They may have to contend with multiple ethnicities, cultures, values, and even languages among their staff and patient populations. They may be frustrated by conflicts between key groups or with entrenched, independent departments or functions, such as their contract pathology and laboratory group or their emergency room group. Administrators may find it difficult to elicit cooperation from constituencies that all seem to have competing financial interests, conflicting agendas, and differing professional loyalties.

The administrative leadership, again whether medically certified or not, faces the chief responsibility of directing, coordinating, monitoring, and reporting on the professional efficacy of teams of free agents, i.e., professionals often not beholden to the delivery institution for their license to serve and at times not accountable to the leadership for the responsibility to serve safely. Administrative leaders thus bear the dual charge of managing independent providers—with their concomitant flows of service, information, and cash—and reporting the net clinical and financial results to their trustees, whether the latter are owners' representatives or agents of the community at large.

When patient safety fails, the trustees turn first to the administrative leadership for insights and remedies: "What happened, and why?" Even when safety hazards recur in the form of hospital-acquired infections in the operating suite, the trustees turn first to the administrator for accurate information and proposed solutions. The inability to provide either may stifle or shorten a promising administrative career.

Competing professional agendas

Several of the key groups upon which patient safety depends often have different expectations. For example:

- Some patients think that a hospital stay should be like a visit to a hotel or spa, not realizing that money spent on fancier rooms or meals could have been spent on making their visits safer.
- Physicians sometimes think that the hospital exists at their pleasure and should cater to their individual ways of doing things, without considering the complexity and increased opportunity for error that this customization creates.
- Administrators, nurses, and physicians often individually think, "We *are* the hospital"—meaning that the essence of the institution is contained in their perspective and presence. Attitudes and behaviors follow that take into account only a limited perspective of the institution.

Professionals are accountable to their profession, and some regard that accountability to be more important than their responsibility to the organization in which they practice. This outlook means that physicians are accountable to their peers. But systems of peer accountability are notoriously ineffective. Doctors policing other doctors does not work very well, not only because it is emotionally difficult to confront a peer but also because physicians can incur legal liability by doing so.[9] Yet administrators and managers may have limited willingness and little power with which to confront and hold physicians and other clinical staff accountable, especially physicians who are on staff but are not employed by the hospital.

Administrators sometimes are faced with midlevel managers (upon whom they must rely) who may have high-level technical skills but few management skills. Too often no one in the organization has the knowledge, skills, and abilities needed to address comprehensively and successfully the organizational issues required to build a strong safety climate. And few come by the required leadership strength naturally.

Many think that the CEO is responsible for fixing the system. Others say there simply *is* no healthcare system. If you consider the extended environment that healthcare encompasses, this claim may be true: healthcare includes hospitals and physician offices, drug and medical equipment manufacturers, community pharmacies, patient homes, emergency transport vehicles, nursing homes, and hospice. Interactions and handoffs among these environments present communication challenges and great opportunities for error, as do the handoffs within the office, hospital, or nursing home. No single person holds a position from which to command and coordinate all these different pieces. It takes leadership and many people—whether clinician or administrator—working together.

Whose job is it to take the lead?

Healthcare safety, as we've said, is complex. The complexity arises from the many constituencies involved and their sometimes competing and conflicting agendas, as well as system and cultural fragmentation, the difficulties of standardizing procedures and practices, inherent professionalism and the lack of hierarchical authority, highly complex technical and technological issues with life-and-death decisions made in crisis situations, plus a litigious atmosphere bearing on people who are often overworked and fatigued.

9 Wachter and Shojania, *Internal Bleeding*, pp. 321–326.

The leader's role is to define the healthcare safety issue for the organization and the terms of the organization's engagement with it. The leader marshals the constituencies that must cooperate to create a strong safety climate and an organizational culture that supports safety. In subsequent chapters we address how this is done well, what it is made up of, and how such efforts are measured.

Throughout this book we refer to the leader of healthcare safety without giving this leader a functional title. This ambiguity stems from our conviction that, when successful, leadership in the matter of patient safety occurs simultaneously at many different levels of the delivery system and within several distinct professional disciplines. One of your challenges in applying these insights will be to discover who else in your organization and on staff needs to join you in taking the lead.

In the next chapter we present a model for understanding healthcare safety that untangles the complexity and arms you with a useful way to think about the issues. It lays the groundwork for understanding exactly what you need to do to lead healthcare safety improvement and build a strong safety climate in your organization.

BLUEPRINT FOR HEALTHCARE

CHAPTER TWO *SAFETY EXCELLENCE*

BLUEPRINT FOR HEALTHCARE SAFETY EXCELLENCE

Many healthcare organizations already have patient safety initiatives under way. Often, these programs are one-off interventions that arose in response to an adverse event or they are an effort to comply with the latest regulatory requirement. Rarely does an organization approach patient safety in a comprehensive, systematic, and disciplined fashion. This is unfortunate both for the safety leader and for the safety effort. It makes leading more difficult and can produce inefficient and haphazard interventions, competing or even conflicting safety components, and confusion among treatment team members and other employees.

Effective safety leaders benefit from a systematic approach to understanding and undertaking safety in their organization. The Blueprint for Healthcare Safety Excellence (Figure 2–1) provides a useful way to sort out the relationship of leadership to organizational culture, safety climate, healthcare safety–enabling elements, organizational sustaining systems, and the working interface and to identify how these components affect the exposure of patients and employees to hazards in the working interface. The blueprint provides a systematic approach to healthcare safety and a conceptual access to effective intervention strategies. The blueprint elements are further explained in Table 2–1.

FIGURE 2-1. BLUEPRINT FOR HEALTHCARE SAFETY EXCELLENCE.

Source: BST analysis

TABLE 2-1. BLUEPRINT ELEMENTS DEFINED.

Leadership	Organizational Culture
Seeing the right thing to do to reach objectives and motivating the organization, whether it is a team or a corporation, to do these things effectively. Safety leadership is exercised by decision making, which is shaped by the beliefs of the leader and demonstrated by his or her behavior.	The driving values of the organization. "The way things are done around here." The unstated and often unconscious assumptions about *how* things are done. (Distinguishable from *safety climate*, which is the prevailing and *immediate* emphasis consciously perceived to be given to patient and employee safety by the organization's leaders.)
Healthcare Safety–Enabling Elements	**Organizational Sustaining Systems**
The set of mechanisms that reduce or eliminate exposures to hazard in the working interface. Different organizations classify these mechanisms in different ways, but they usually include hazard recognition and mitigation, incident root cause analysis, training, regulations, procedures, policies, and safety improvement programs.	The set of systems that maintains the healthcare safety–enabling elements and ensures their effectiveness. The systems include organizational structure, selection and development of people, performance management, financial rewards, and degree of employee engagement.
Working Interface	
The environment in which exposure to hazards is created, mitigated, or eliminated, and where adverse events occur or are prevented. Not just the bedside and other clinical settings but also the laboratory, magnetic resonance imaging unit, hallway, patient shuttles, etc.—any place staff, patients, and/or technology interact.	

Organizational safety is embedded in a matrix of forces beyond the local healthcare organization: the national culture, the demands of the marketplace and of third-party payers, the tort system, and professional and local standards of practice. These forces all work through the various elements of the blueprint to render their consequences in the working interface. Together, these forces and the elements depicted in the blueprint encompass most or all the important variables bearing on healthcare safety.

The various elements depicted in the blueprint capture the immediate organizational causes of patient safety. If an adverse event occurs, its root causes doubtless will be found among them. Since the blueprint encompasses all the local patient safety variables, it provides a template with which to assess the current state of the organization and to define the components of a safety vision.

We know that leadership is important to safety excellence, but what are the specific mechanisms that connect board members, the CEO, the physician, the chief pharmacist, and the head nurse to safety performance improvement? What specific mechanisms connect you to the results you seek? By "mechanism" we mean a set of steps or system components that reliably lead to a defined result.[1] It is not sufficient to say that management needs to support the safety effort. Of course management support is needed, but we need to be much more specific. What do you need to understand and do to ensure success? What common threads run through safety improvement mechanisms, and what behaviors are required to ensure them? How do the things you do and say, or fail to do and say, influence the working configuration and the way treatment is administered to patients?

As we explore the elements of safety depicted in the blueprint:

- Consider how your organization would score on an assessment of each of these elements.
- Delineate your role as a safety leader in terms of each element.
- Begin formulating your vision for safety in terms of each of the elements.

Although the examples in this book usually describe patient safety issues, the core concepts apply equally to employee safety. It will help you to become fluent in the use of these concepts if, for every patient safety example we offer, you think of a corresponding employee safety example as well.

The working interface: Where exposure to hazard can occur

The working interface is the configuration of equipment, facilities, processes, systems, and behaviors that define the interactions among patients, staff, and technology. The working interface can

1 Thomas R. Krause, *Leading with Safety* (Hoboken, NJ: John Wiley & Sons, 2005), p. 147.

be characterized as exposing the patient to some average number of hazards per unit of time and thereby producing some number of adverse events. If an adverse event occurs and a patient is harmed, the working interface is where it happens.

By "exposure to hazard" we mean any condition, decision, behavior, activity, cultural standard, process, or system (or lack thereof) that increases the probability of the patient's suffering a preventable adverse event. In short, any malfunction in any element of the blueprint that increases the patient's jeopardy constitutes an exposure and is a root cause of preventable adverse events. By "preventable adverse event" we mean healthcare-caused harm, i.e., a patient injury that is not an inevitable or necessary outcome of the patient's illness but rather the result of the care he or she received. Table 2–2 shows some common healthcare safety exposures and related preventable adverse events.

TABLE 2-2. EXPOSURES AND PREVENTABLE ADVERSE EVENTS.

EXPOSURE	PREVENTABLE ADVERSE EVENT
Understaffing	Inability to follow procedures
Failure to identify the patient with whom you are interacting	Performing an operative procedure on the wrong patient
Failure to administer medications on the schedule ordered	Treatment failure
Failure to comply with hygiene best practices	Nosocomial infection

Relationship between exposure and preventable adverse events

H. W. Heinrich first described the relationship between exposure and preventable adverse events in 1959,[2] and it has been emphasized in most standard safety texts ever since. It is expressed as the familiar safety pyramid (Figure 2–2), which illustrates two points:

- Exposures far exceed adverse events. Accordingly, the most likely outcome of an exposure is not an adverse

2 H. W. Heinrich, *Industrial Accident Prevention: A Scientific Approach*, 4th ed. (New York: McGraw Hill, 1959).

event; rather, the most likely outcome is that nothing harmful happens. This observation sheds light on why hazardous shortcuts naturally proliferate and are often not self-correcting.

- There is an inverse relationship between frequency and severity: for every serious event, there are many less serious events. Resistance to this insight—demonstrated again and again in both clinical and industrial settings—often explains the failure of organizations to reduce the frequency of exposures to hazard and the resultant injury rates.

FIGURE 2-2. RELATIONSHIP BETWEEN EXPOSURE AND PREVENTABLE ADVERSE EVENTS.

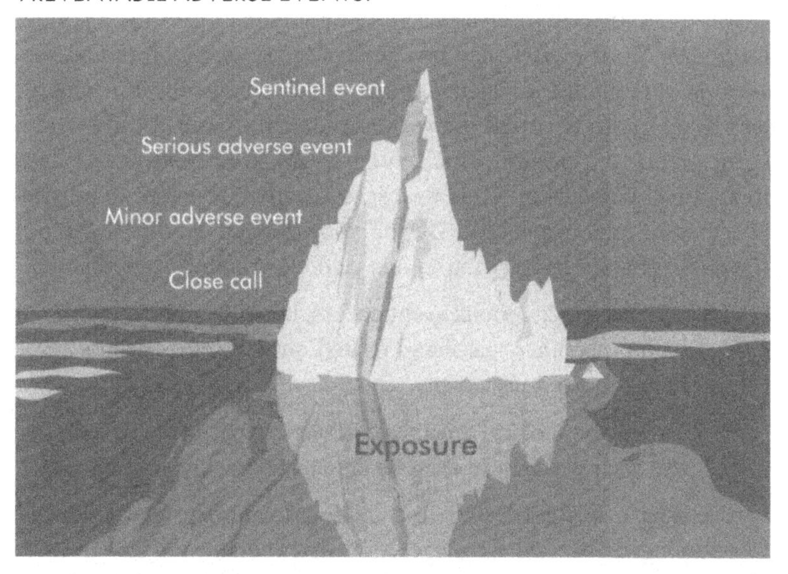

The idea that there is an inverse relationship between frequency and severity has been criticized[3,4] because it does not account for the great disparity between the likelihood of various hazards causing serious injuries. Critics argue that when the focus is on frequency alone, intervention efforts can be

---

3 D. Kriebel, "Occupational Injuries: Factors Associated with Frequency and Severity," *International Archive of Occupational and Environmental Health,* 50 (1982): pp. 209–218.

4 Fred A. Manuele, *Heinrich Revisited: Truisms or Myths* (Itasca, NY: National Safety Council Press, 2002).

placed mistakenly on exposures associated with less serious adverse events. This criticism is valid. Of course, not all exposures are equal in terms of the potential for adverse events.[5] Some exposures result in more serious incidents; some in less serious ones.

Nevertheless, an environment that frequently generates low-severity events—be they inconsequential medical errors or minor employee injuries—harbors systems, cultural, and leadership issues that can (and eventually will) generate high-severity events as well. It would be a serious mistake to disregard the fact that frequent low-severity events indicate the potential for high-severity events. Certain kinds of low-severity events are distinguished from sentinel events only by good fortune.

The underlying principle is that where common causes are involved, many small or less severe events may presage a single large or serious one. Those smaller or less severe events may be similar in type but lower in severity; e.g., the wrong but harmless drug given to a patient versus the wrong and highly toxic drug given to a patient. Such events may also be earlier links in a causal chain that leads to a serious adverse event; e.g., the practice of placing look-alike or sound-alike medications on the same shelf. This principle has two significant implications:

- When a single serious event occurs, it can be inferred with high probability that many related but less severe events have occurred previously.
- To prevent medical errors and adverse events, small events and their precursors must be taken as seriously as large ones.

Consider a student living in a college dormitory who is found to be using heroin. The college administration wants to know if this is an isolated event, or if other students are also using the drug. Our principle suggests that the probability of a single serious event occurring in isolation is very low and that if one serious event has occurred, it is highly likely that many other less serious events of the same kind have happened previously. In this case, one student using heroin indicates that other students on campus are probably also using drugs.

5 Fred A. Manuele, "Injury Ratios: An Alternative Approach for Safety Professionals," *Professional Safety*, 49 (February 2004): pp. 22–30.

Similarly, if a single preventable patient fatality occurs, it is highly likely that its causal roots—the particular exposures that caused it—occur frequently. This is not to say these exposures are constantly present. Their occurrence is likely to be intermittent and at unpredictable intervals. Nevertheless, each time a patient is exposed to a hazard, the exposure represents an important risk; no matter that no patient harm has yet resulted. Thus, an injury-free culture is not simply one that doesn't tolerate incidents; it is one that doesn't tolerate exposure to hazards.

Leaders are often slow to recognize the statistical facts of the situation. It is easier—almost natural—to think that the rare seri-

An injury-free culture is not simply one that doesn't tolerate incidents; it is one that doesn't tolerate exposure to hazards.

ous event is a fluke. As our statistics professor used to say, "Rare events happen, rarely." Serious adverse events may happen rarely, but their causal roots—the working interface hazards that lead to serious events—are far more frequent, and their presence virtually guarantees that a serious event eventually will happen, too.

Imagine that this morning on the way to the hospital you observe an automobile accident. A car swerves into oncoming traffic. The driver is not wearing a seat belt and is thrown from the car and killed. What is the probability that this driver usually wore a seat belt but just didn't happen to be wearing one today? It could happen, but the likelihood is low. It is more likely that the driver frequently did not wear a seat belt, and today was no exception. And how many years had he been driving without a seat belt before the odds caught up with him? Probably a long time. If a friend had told him he ought to wear his seat belt, he probably thought he didn't need one: he had been driving for years without an incident.

Similarly, let's imagine that a patient dies of a heparin overdose and the incident investigation finds that the heparin for her IV flush had been kept at the bedside along with her IV push medications. What is the likelihood that keeping IV meds at the bedside is a rare event on that unit? It is possible, but the likelihood is low. It is more likely that this exposure occurs with some regularity because of its convenience and expedience, and because "nothing bad has ever happened." It has probably produced unrecognized or unreported near misses or minor injuries in the past.[6] In hindsight we can see that the major event was inevitable; it was only a matter of time.

Infrequency of adverse events leads to resistance

One of the most common resistances to tackling patient safety is voiced by physicians and administrators alike. It is the claim, "We don't have a patient safety problem here. We haven't had a serious event in *x* period!" (fill in the time frame of your choice).

Rare serious events are not flukes. They float on a sea of unacceptable exposures. A leader—whether clinical or administrative—can legitimately claim not to have a patient safety problem *only* on the basis of a concrete and replicable measure of exposure. Often, however, those who claim not to have a problem base their claim explicitly on the rarity of events and implicitly on their ignorance of exposure.

To senior administrators fall the responsibility of knowing the frequency of exposure and the means of mitigating hazard. They also hold leading accountability for enrolling their clinical "free agents" in the necessary changes of behavior and attitude that will reduce or eliminate exposures to hazard.

Frequency of exposure as part of the culture

The IV heparin overdose example illustrates another important point: if, on this unit, IV push medications are frequently kept at bedside, it means that this practice has become a part of the unit's culture. And that means that local unit leadership has tacitly accepted the hazard. If an exposure occurs with some regularity, it becomes a part of the culture and is regarded by those laboring at the bedside as accepted by the leadership.

---

6 The near misses and minor injuries represent the first instance of so-called upstream events in this text. The noticing, recording, and proactive response to these minor events upstream of a serious adverse event are critical to the avoidance of major, preventable incidents.

The same dynamic is at work in the seat belt example. The longer the driver goes without a seat belt, the more natural it feels to him. One can see this same dynamic gradually being reversed as our culture begins to reject secondhand smoke and the poor nutrition of fast foods—exposures that used to be taken for granted or considered harmless.

Statistical variability and leadership overreaction

A given number of exposures in a period can engender a wide range in the number of adverse events in the same period. A given exposure will have a different result today from what it might have tomorrow, simply by chance. If this basic statistical relationship is not understood, leaders will inevitably overreact to adverse event data. Some months will go by in which adverse event frequency is unusually low, and leaders will conclude that patient safety is improving. In fact, it may or may not be improving. Or in a period of a few months a rash of adverse events will occur, and leaders will say that patient safety has deteriorated. In fact, exposure may have been reduced during this period.

None of this is to say that safety is ultimately a matter of luck, but the random variability in adverse event frequency means that effective interventions should be triggered by exposures—which are far less subject to random variability—not just by serious adverse events. Even more than major adverse events, exposures signal the need for action, and major adverse events signal that many exposures have been discounted or overlooked.

The use of safety leading indicators

The random variability in adverse event frequency means that, to lead effectively, a leader requires leading indicators. Leading indicators are measures of variables that can be shown to have a statistically valid, predictive relationship to adverse event frequency. When viewed in relation to lagging indicators (for example, the number of adverse events in a period of time), leading indicators allow organizations to take proactive steps to prevent patient injuries. Table 2–3 shows some healthcare safety leading indicators.

TABLE 2-3. HEALTHCARE SAFETY LEADING INDICATORS.

Working interface assessment (percent safe practices). For example, frequency of compliance with best practices, such as the following:[7] • Appropriate use of prophylaxis to prevent venous thromboembolism • Use of perioperative beta blockers in appropriate patients • Use of maximum sterile barriers while placing intravenous catheters • Appropriate use of preoperative prophylactic antibiotics • Requiring patients to restate what they have been told during informed consent • Continuous aspiration in ventilated patients • Use of pressure-relieving bedding materials • Use of real-time ultrasound guidance during central line insertion • Training patients in self-management for warfarin • Appropriate provision of nutrition, with a particular emphasis on early enteral nutrition in critically ill and surgical patients
Extent of training on policies, procedures, and protocols
Thoroughness of communication to clinical teams regarding protocol requirements
Frequency of physician reassessment of patient goals upon readmission
Role-play of critical behaviors (percent satisfactory performance)
Critical behavior observation (percent safe)
Rate of critical policy, procedure, process compliance (e.g., use of written tools to obtain truly informed consent)
Near-miss reporting and rate
Caregiver and administrative proactive engagement (percent participation)

One of the most surprising aspects of this distinction between leading and lagging safety indicators is that most safety professionals already understand it, while organizational leaders, including executives who make important safety-related decisions, frequently do not.

7 Kaveh G. Shojania et al., eds. *Making Healthcare Safer: A Critical Analysis of Patient Safety Practices* (Rockville, MD: Agency for Healthcare Research and Quality, 2001). AHRQ Publication No. 01-E058. Viewed at www.ahrq.gov/clinic/ptsafety/

Controlling exposure in the working interface

Excellence in patient safety is directly related to how effectively the organization controls exposure to hazards in the working interface. Each of the other elements of the blueprint plays a critical role in controlling this exposure, and one's effectiveness as a safety leader is a function of the ability to influence these elements. Thus, the healthcare leader's safety role can be understood to be that of leading the diagnosis and treatment of all those parts of the blueprint that need attention and that, when given attention, promise further reduction of exposures.

By implication, to control exposures the administrative leadership holds multiple responsibilities:

- Engaging both clinical and administrative staff in the identification of exposures at the bedside
- Gathering concrete and verifiable data regarding exposures
- Developing leading indicators that enable the mitigation or elimination of exposures
- Informing both hospital-based and independent clinicians of the importance of eliminating exposures
- Finding specific ways to mitigate hazards
- Reporting the ongoing frequency of hazards to the board

Healthcare's extended working interface

We sometimes use the phrase "the bedside" to refer to the working interface. Actually, healthcare's working interface extends far beyond the bedside in both space and time. It is where any aspect of treatment occurs or wherever healthcare workers do their jobs. Thus it includes not only the hospital bedside, clinic exam room, treatment room, and laboratory but also medical transportation, patient homes, extended care facilities, phone conversations between patients and nurses, hospice, courier vans transporting laboratory specimens to laboratories, and community pharmacies. (The working interface even includes so-called curbside encounters, where the patient runs into his physician on the street and says, "Hey, Doc—I stopped taking that medicine you prescribed because I don't like the way it makes me feel. Okay by you?" Because there's no medical record in front of the physician and no opportunity to make an extemporaneous notation on the patient chart, interactions like this can be particularly hazardous.)

That healthcare's working interface comprises so many venues and time periods is a frequent source of medical errors: it necessitates the effective, often complex, and regularly difficult communication between and among professionals in the various venues. One physician confided that his greatest fear is that a patient will be harmed because his hospital discharge orders have not been adequately communicated to those responsible for care delivery beyond the hospital walls.

One of the authors can personally testify to the communication problem. After a brief hospital stay, he found he was unable to remember the simple instructions he had been given at discharge. If he, a physician familiar with all the medications and procedures involved, had difficulty recalling the instructions, imagine the difficulty of an elderly, infirm, or cognitively compromised patient. It would have helped him greatly to have received the instructions in writing as well as verbally. This simple intervention would have enhanced a safety-enabling element in the working interface.

Healthcare safety-enabling elements

Safety-enabling elements are all those mechanisms specifically introduced into the working interface to reduce or eliminate exposures to hazard. These are the basic elements of safety, including policies, procedures, accreditation, knowledge, skills, hazard recognition, incident root cause analysis, and specific safety programs. In industry, most large organizations have these programs in place and audit them regularly. In healthcare, this practice is not as frequently observed. Which of these safety elements are needed in your organization? Which are in place? How user efficient and friendly are they? How are they audited, and how effective are they?

Recent healthcare safety–enabling improvements include:

- Instituting computerized physician order entry to eliminate physician handwriting and medication interaction errors
- Signing the surgical site to eliminate wrong-site surgery
- Using a morbidity and mortality conference to surface errors and conduct root cause analyses

Sometimes a safety-enabling element will have unanticipated or unappreciated negative consequences as well as the salutary ones intended. Restraining elderly nursing home residents who have been particularly unsteady on their feet is an enabling element designed to protect them from falling. But restraint has many deleterious consequences, not the least of which is making the patient weaker and thus less steady. Some nursing homes have instituted exercise and movement regimens to improve strength and balance.

Most organizations have discovered that two locations can have practically identical audit scores on safety-enabling elements, identical or near-identical technology, and similar workforces, and yet report widely different incident frequency rates.[8] Enabling elements are necessary but not sufficient for excellent safety performance. As we will see, there is a great deal more to safety success.

Organizational sustaining systems

Organizational sustaining systems include all business systems that maintain and ensure the effectiveness of safety-enabling elements across time. Although most organizations have these systems, their quality varies substantially. Perhaps more important, most organizations fail to appreciate and attend to the relationship of these systems to safety-enabling elements, to the working interface, and to patient safety outcomes.

8 Dan Petersen, "The Four Cs of Safety: Culture, Competency, Consequences and Continuous Improvement," *Professional Safety* (April 1998), pp. 32–34

The systems that are necessary to sustain excellent performance include:

- Hiring and training functions
- Performance management systems, including processes for defining individual performance expectations and tracking personal performance
- Development and succession planning
- Conscious assignment of decision latitude to align talent with risks of exposure
- Audits of safety-enabling elements
- Systems to generate and use leading indicator data
- Regular board-level review of leading and lagging safety indicators
- Regular observation and review of the degree of employee engagement in safety

So, for example, in your organization is safety leadership a criterion for jobs that are central to safety outcomes? Is there a process to develop patient safety leaders? Is the structure of the organization such that patient safety is given adequate emphasis? Does the performance management system meaningfully address safety leadership issues—not just through lagging indicators? Are there mechanisms to ensure employee engagement in safety? Is there a systematic way to hold leaders accountable for safety processes and outcomes?

For brevity's sake we sometimes use "systems" to refer both to safety-enabling elements and to organizational sustaining systems. Much current patient safety activity is focused at the level of these systems. Our (unpublished) survey of the May 2006 activity in the Patient Safety Discussion Forum, an Internet newsgroup[9] sponsored by the National Patient Safety Foundation, revealed that almost 90% of the forum's discussion threads dealt with systems issues. The task of addressing systems issues can be safely delegated[10] provided the

9 Available since February 1999 at http://listserv.npsf.org/.

10 To whom they should be delegated is a tactical implementation decision, which we will discuss in chapter 9.

organization's culture has a strong value for safety. If it does not, then delegation runs the risk of fragmenting efforts, aggravating existing turf warfare, increasing the tendency to blame, and ultimately engendering a failure to sustain hard-won improvements.

Some of the systems commonly involved or found lacking in preventable adverse events include the following:

- Patient handoff mechanisms and systems that manage communication between various caregivers, services (laboratory, radiology, etc.), and working interface venues
- Physician order entry system and systems for medication storage, administration, and reconciliation
- Systems for patient identification and tracking
- Systems to plan for and respond to patient flow and progression issues
- Systems to ensure adequate assessment of applicants and to qualify new hires
- Systems to ensure that critical skills are maintained at a high level (e.g., Code Blue response team drills)

In addition, processes to ensure that findings from the root cause analysis of incidents are effectively used to prevent future incidents are also critical but often weak. They are weak because, although management organizations typically assign clear responsibilities for the initiation, ratification, implementation, and monitoring of major initiatives and investments, they rarely assign similarly clear duties for the notification, review, and adjudication of analytical findings, and they even more rarely do a good job of following up on whether implemented solutions are doing what they were intended to do. Because of the number and prominence of independent clinicians in healthcare delivery organizations, these information roles are highly consequential. The administrative leadership needs to spread the word when many members of the audience may have only a passing interest in hearing it.

Both enabling and sustaining systems receive heavy focus in repeated cycles of "strategic" initiatives meant to improve healthcare, among them Total Quality Management, Six Sigma, Lean, right-sizing, world-class service, electronic medical records, capitation, hospitals without walls, and supply chain realignment. While each such initiative holds its own promise of improvement in the quality and efficacy of care, most miss a huge opportunity to focus on organizational culture and induce productive changes in each system, whether sustaining or enabling. Focusing on the systems themselves represents an essentially tactical approach to healthcare safety. It may miss the strategic advantages of a focus on culture.

Specifically, although safety-enabling elements and organizational sustaining systems are unquestionably important to a safe working interface, the strength of them alone usually does not suffice to predict (or control) variability in performance. Rather, differences in organizational culture and climate illuminate and explain the variations in the frequency of preventable adverse events between otherwise similar organizations. If two hospitals, for example, have similarly well-developed healthcare safety–enabling elements and organizational sustaining systems, and similar technologies, staffs, and patient populations but different frequency rates of patient injury, the difference between them is likely to be found in their cultures.

Organizational culture

When healthcare professionals hear the terms *culture* and *safety climate,* they often sound too vague to be helpful. In this book we mean something specific and measurable by these terms, and we know from our experience and from published social science research which aspects of organizational culture and safety climate are critical to safety excellence (discussed at length in the next chapter).

Culture versus climate

Culture and climate are related but distinct concepts:

- *Organizational culture* means the shared, often unconscious values, attitudes, standards, and assumptions that govern behavior, especially in situations that lack clearly defined rules and procedures. Organizational culture is the driving values of the organization—"the way things are done around here." Culture can be codified in concrete and quantifiable terms that lend themselves to measurement and intervention.

- *Safety climate* refers to the prevailing, usually consciously held perception of leadership's priority for a particular area of functioning (patient and employee safety in our case) at a particular time. It reveals to workers with immediacy and urgency how leaders want them to respond in the face of conflicting priorities.

Thus organizational culture is deeply embedded and long term; it takes longer to change and drives organizational performance across many areas of functioning. Safety climate, on the other hand, changes faster and more immediately. It reflects the prevailing attention of leadership.

Think of organizational culture as an influence on the organization that operates powerfully but unseen in the background. Safety climate, on the other hand, operates in the foreground. It provides immediate and prominent signals regarding what is wanted and needed in the moment. Climate changes faster than culture, and *a climate change, when sustained, reshapes culture over time.* The practical importance of this distinction is that, while culture accumulates from the words, thoughts, and actions of leaders over time—and is therefore relatively inaccessible to immediate leadership intervention—climate is immediately accessible. If you sustain and reinforce an intentional change in climate, you drive a change in culture.

Culture, safety climate, and the working interface

To appreciate the profound importance of culture and climate to the safety of the working interface, consider that medical errors most frequently occur in situations in which providers must deviate from procedure or in which procedures are weakly defined or nonexistent. It is in these situations that culture comes most strongly into play to define how well things are done. Think, for example, of the story from chapter 1 about the loss of life resulting from two open containers of unidentified fluids in the operating room. The culture apparently (again, we are speculating in order to make a point) did not provide any guidance to make sure that the anesthesiologist was using the correct substance, or that the nurse and anesthesiologist communicated about what each was doing. He didn't ask her, "Is this what I think it is? Am I getting what I asked for? And did I indeed ask for the right thing?" She didn't offer the information, perhaps because speaking up was generally not rewarded. Neither team member labeled the containers they were working with. Instead, each tended to his or her own separate tasks—again, in good faith.

It is usually the so-called little things that hurt us. Our best protection against these hazards lies in a culture that strongly values safety and a safety climate operating vividly in the foreground. Consider the commitment of the leadership at Kennestone Hospital in Marietta, Georgia, to reducing hospital-acquired infections (HAI). Virtually every room and every hallway contains one to several dispensers of liquid hand disinfectant, and senior physicians can be seen cleansing their hands again and again as they move from room to room. The climate of HAI avoidance is vivid and tangible.

In the absence of explicit rules and procedures, organizational culture and safety climate subtly but profoundly guide decision making and behavior. For example, in an organization with a strong safety climate, a junior staff member will choose to speak up if she suspects an error is being made by a senior person. She will speak up because of the longstanding history in the organization that such "bad" news is welcomed as a hazard mitigated for the future. This venturesome step may not happen in an organization with a weaker safety climate. And in situations where there are explicit rules and procedures, culture profoundly influences whether and how the procedures are carried out; e.g., whether a mandatory sponge count is thoroughly or perfunctorily performed, or is even skipped.

What are the characteristics of your organization's culture that most support safety? Where are its challenges? What messages about your priorities define your current safety climate?

The charge of the safety leader

Leadership is about seeing the right things to do to reach organizational objectives and then motivating the organization, be it a team or a institution, to do these things properly and effectively. It is manifested by decision making, which is related to the beliefs of the leader and demonstrated by his or her behavior. In short, whereas management has to do with getting things done, leadership has to do with selecting what to do and knowing how to do it. Administrative leadership, whether clinically certified or not, bears the charge for undertaking both leadership and management on a continuing basis.

Leadership and culture determine how well safety-enabling elements and sustaining systems function, and, as we have said, under some circumstances the task of addressing systems issues can be successfully delegated to others. The same is not true for building a culture that truly values safety. Safety-enabling elements and sustaining systems are managed, but the creation of organizational culture and safety climate is led. Leadership and culture building cannot be delegated.

Thus, although it may be safe to rely on your patient safety officer for tactical systems improvements, it is rarely successful to rely on him or her to lead culture change. So, given the complexity of the healthcare world, how does one build a culture that values safety?

We will have a lot more to say about this in subsequent chapters; suffice it here to say that leaders create organizational culture. For example, how one responds to systems issues—the approach you take and the decisions you make—directly shapes the organization's safety climate. One of the issues facing leaders who desire to change their organization's culture is how best to make their commitment to change known. Problems in safety-enabling elements or organizational sustaining systems provide valuable culture change opportunities. Use them to make your underlying ethical commitments to patient safety visible and tangible.

While the ultimate safety improvement objective is to discharge patients who have experienced no preventable adverse events, the proximate objective is to create a strong, consistent safety climate and a culture in which safety (and the avoidance of exposure to unneeded hazards) is a driving value. But if we look realistically at our healthcare delivery organizations, we often see low trust, poor communication, and deficient leadership credibility.

Many leaders fail to address exposures in the known working interface, and the professional caregivers—nurses, doctors, pharmacists, and others—are often not actively engaged in patient safety. How does leadership change this? How do you create a culture in which safety really is a driving value?

The culture change process starts with you. You and a core group of other leaders need to align with one another on what you value and, specifically, on the value patient safety has for you:

- Understand what principles represent its value.
- Determine what behaviors are necessary to convey to the organization and its professionals how serious you are about patient safety and the culture change it entails.
- Request and acknowledge the right behaviors among others.
- Create consistency among leaders: making the right decisions, communicating the right information, and articulating the right vision.

Culture changes slowly, but it is changing all the time. Leaders are always changing the culture. Every time you make a decision, leave an issue hanging, take a stand, or address an issue, your behavior leaves a mark on the culture. Successful organizational change is about directing and accelerating this natural process of change.

That it takes time to change a culture is both good and bad. The bad news is that a weak or ineffective culture may be resistant to change. The good news is that when leadership makes safety a driving value in the organizational culture, the value endures.

Avoid a culture of blame

Many in the safety community believe a dominant majority of incidents result from behavioral causes, while the remainder relate to equipment and facilities (Figure 2–3). We thought so, too, at the time of our first book, published in 1990. However, we now recognize that this dichotomy, while ingrained in large segments of the safety community, is not useful, and in fact can be harmful.

The either/or choice depicted does not accurately reflect what causes injuries. The equipment doesn't simply malfunction, independently of how it has been designed, maintained, and used, and the healthcare provider doesn't simply behave unsafely, independently of the system configuration. Rather, the provider interacts with the patient and the technology, and the interaction that results constitutes a system.

FIGURE 2-3. WHAT'S WRONG WITH THIS PICTURE?

The notion that most accidents are caused by unsafe acts creates a dichotomy between the action of the worker and the facility, equipment, and conditions of the work—and produces counterproductive blaming.

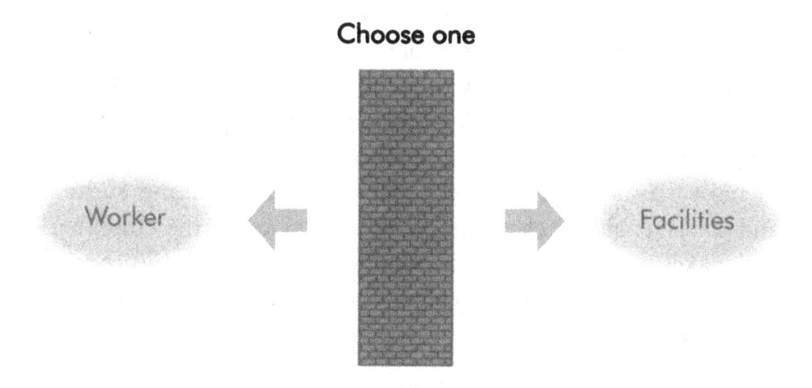

Choosing between human causes and other causes encourages blaming. If the purpose of understanding what causes injuries is to establish fault and win malpractice awards, it is useful to blame the healthcare service providers and the institutions in which they work. But if the purpose is to prevent future injuries, blaming is counterproductive. A culture of blame is also destructive. To the original injury to the patient, a blaming culture can add the insult of inducing the investigation committee to waste time, make poor recommendations, and subvert the safety climate by shielding fellow providers from blame. In the process, careers can be destroyed.

If the objective is to improve patient safety, we must rein in the tendency to blame and—the other side of that same coin—the tendency to look the other way or to cover up. Instead we must aim for a culture that is transparent and seeks to understand how best to improve critical systems, team functioning, the organization's safety climate, and safety leadership. In many organizations, this aim requires major changes in the culture.

Responsibility and accountability do not mean blame. Rather, leadership must create a climate and culture in which it is safe for people to participate wholeheartedly and candidly in adverse event investigations and improvement efforts. The useful question is not, "Who was at fault?" or "What is at fault?" but rather, "How can this event, and others like it, be prevented in the future?" And, most important, "What can I do to help?"

Psychologist James Reason[11] of the University of Manchester uses a helpful metaphor to discuss the problem of blame. He compares human endeavors to a chisel (Figure 2–4) with two ends:[12] the sharp end, representing in our terms the healthcare delivery providers interacting directly with the patient in the working interface, and the blunt end, which receives the force of the mallet striking it, representing in our terms all the forces impinging on the working interface and influencing how the providers and patients behave, including leadership, culture, organizational sustaining systems, and safety-enabling elements. Who is to blame if the chisel slips and makes a wrong cut? Is it leadership, systems, and culture at the blunt end, or is it (as more typically assumed) the providers at the sharp end?

---

11 Jim Reason has written books on absent-mindedness, human error, aviation human factors, managing the risks of organizational accidents and managing maintenance error. Since 2000 his primary focus has been patient safety. Viewed at www.saferhealthcare.org.uk.

12 Robert M. Wachter and Kaveh G. Shojania, *Internal Bleeding: The Truth Behind America's Epidemic of Medical Mistakes* (New York: Rugged Land, 2005), p. 43.

FIGURE 2-4. WHICH END OF THE CHISEL SHOULD BE BLAMED FOR THE ERROR?

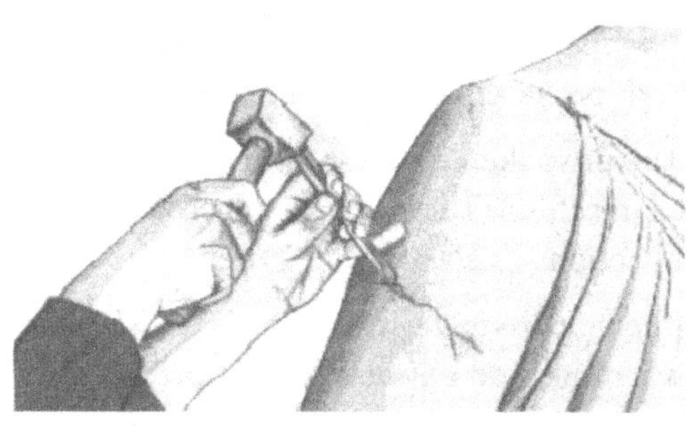

Clearly, this question assumes a simple-minded view of injury causation. The chisel's sharp and blunt ends are inseparable. When adverse events occur, there are critical interactions between the people in the working interface and the other elements of the blueprint. It is these interactions that we must understand. How are leadership, systems, and culture impinging on treatment team members and patients in ways that create hazards? Getting safety right means systematically influencing these interactions to reduce or eliminate exposure.

This improvement requires the cooperation of everyone involved. An implication of Reason's chisel is that no one can legitimately claim to be uninvolved in patient safety. "It's not my job" doesn't wash. It is not legitimate for an administrator to feel that patient safety is the responsibility of the nurses and doctors; nor is it legitimate for a doctor or nurse to claim that he's too busy and that responsibility for systems issues belongs to the administration.

Analyze adverse events

Consider once again the case from chapter 1 in which an anesthesiologist injected glutaraldehyde into a patient's spinal canal from an unlabeled container. One might think that the nurse was to blame for this terrible accident. After all, she placed the poisonous substance on the sterile field. Or perhaps the anesthesiologist was to blame; why didn't he ask instead of assuming? But the analysis of the event from the point of view of the Blueprint for Healthcare Safety Excellence enables us to ask the questions that need to be asked to develop a comprehensive understanding of the event.

Organizational culture

- Is the character of the operating room culture such that team members describe critical actions to other team members as the actions are being performed?
- Does the unspoken understanding between nurse and physician require that the nurse verbally call out her critical actions as she executes them?
- Does the culture foster the nurse's willingness to openly give warnings and express her concerns?
- Does the anesthesiologist routinely lead the operating room team in a discussion of anything unusual about the procedure before the operation begins, clarifying who will be doing what?

Organizational sustaining systems

- Have the anesthesiologist and operating room nurse received training on the various steps in the surgical procedure, and are written instructions available for them to study before the operation?
- Does the quality improvement system perform analyses on rarely performed critical procedures?

Safety-enabling elements

- Is there a procedure requiring the anesthesiologist to confirm the contents of every container of fluid before its use?
- Is there a procedure for labeling fluids the moment they are collected (spinal fluid) or poured (glutaraldehyde)?
- Is there a procedure for differentiating by container color those fluids that are hazardous?

Leadership

- Are leadership roles and responsibilities in the operating room clearly delineated?
- Are organizational leaders systematically and promptly informed of every adverse event?

- Are patient safety leaders acting responsibly to proactively shape the climate and ensure systems are in place and functioning properly to prevent such events?
- Do they have the data they need to do so effectively?

If we fail to think in this way about adverse events, we focus on what happened at the sharp end, fire the nurse, and summon our lawyers and loss prevention officers. We forfeit the opportunity to figure out how to fix the problems at the chisel's blunt end.

Blame is rarely an appropriate response to a medical error. Of course there are rare situations where blame is appropriate: for example, criminal behavior, insubordination, or refusal to respond to feedback. But these behaviors are rare and seldom the cause of preventable adverse events. On the other hand, a culture of blame is an exceedingly common problem in healthcare and frequently contributes to adverse events by impeding detection, investigation, and remedy. Every effort should be made to eliminate blame from healthcare's culture.

A pathologist attending a medmal presentation heard about the case of the patient who died from an injection of glutaraldehyde and became worried. He went back to his own hospital, where he was medical director of the laboratory, and at the next meeting of the surgical care committee, of which he was part, he asked about procedures in the operating room with regard to glutaraldehyde and formaldehyde. The surgeons told him, "We have big bottles of the stuff, and we just pour out what we need when we need it."

The pathologist proposed a new rule: no more open bottles of preservative in the operating room. Instead, his laboratory would prefill, seal, and label (with skull and crossbones symbols, prominent "poison" warnings, and expiration date) various sized plastic specimen containers and stock them in the operating room. The laboratory would supply all the standard sizes (tall, short, large, small), and when a surgeon needed an unusual size (e.g., a container large enough for a placenta or kidney), the laborabory would immediately prepare a customized container, close it, label it, and bring it to the operating room.

The surgical care committee, made up of surgeons, balked at the pathologist's suggestion. "Why are you trying to tell us how to run our OR?" they demanded. "We have this big jug—we don't want to waste it. Besides, your idea will take too much time." The pathologist convinced them that the new system would actually save time, because no one in the operating room would have to do any pouring. Most important, patient lives were at stake.

Once the pathologist got the cooperation of the surgical review committee, he carried the proposal to the hospital's medical executive committee, which includes hospital administration. Either the vice president or president always sits in. The executive committee saw the truth and value of the idea and directed the nurse who was the head of the operating room: "Henceforth you will do it this way. Use the lab's prefilled, labeled containers. Waste all the preservative you need to. There will be no more pouring of preservative in the operating room." In this way a new safety-enabling system was born.

Safety improves when someone becomes aware of a hazard and takes action to remove it—by creating a new system, if necessary. But for a safety-enabling system to work, safety needs to be the *easiest* thing to do, not the hardest. The way to change systems, particularly in organizations with rigid hierarchies, is to make the *right* way the *easiest* way. Nowadays prefilled specimen containers with preservative are commercially available, not because of that pathologist but because others, over time, also recognized the hazard, and the healthcare products industry responded to the need.

The new system, as good as it is, would not in and of itself address all the cultural and communication issues that may have coalesced to create the hazard, the error, and the patient's death. But its creation is an example of effective safety leadership at work.

Acquire the skills

How effectively a person fulfills the safety leader's role—leading the effort to improve his or her organization in all aspects of the Blueprint for Healthcare Safety Excellence—is a function of the leader's skills, knowledge, and abilities, but also, and especially, of the value the leader places on patient safety. And of course, people vary. This is as true for the CEO as for the nurse at the bedside. Some leaders have a natural inclination toward safety and need little help; others are quite reluctant to take on safety issues and may even be apprehensive about it. Leaders are often chosen for their technical ability, and providing excellent safety leadership is necessarily a people activity requiring high levels of interpersonal skills.

If you feel you lack some aspect of the leadership skills, knowledge, or abilities you need, you need not worry. Knowledge can be learned and skills cultivated. The important things are that you have compassion for patients, care about their safety, are willing to act, and will persevere in your efforts. When these capabilities are active in an organization, they are manifested in obvious ways (to be discussed in chapter 4). Great safety leaders are simply great

Great safety leaders are simply great leaders who are motivated to improve safety.

leaders who are motivated to improve safety; they are no different from great leaders generally.

Although the strength of safety leadership and culture varies from organization to organization, we know what strength in safety leadership consists of and what a strong safety climate looks like—and we can help you get there:

- We can define the attributes of a healthy organizational culture and safety climate and specify how to measure them (to be discussed in chapter 3).
- We know the values, leadership styles, and best practices of great safety leaders. We can describe the specific behaviors and practices they employ to provide safety leadership and provoke culture change (to be discussed in chapters 4 and 5).
- We know safety leadership behaviors are subject to the same principles as other behaviors. Understanding these principles makes it possible to quickly acquire the behaviors needed to lead safety and to build a strong safety climate.

- We know that a vision of safety leadership provides a path toward excellence that anyone who aspires to excellence in patient safety can travel.

In this chapter we learned that providers are not to blame for the adverse events that occur in the working interface. Rather, these events are the result of complex interactions among the many elements of the Blueprint for Healthcare Safety Excellence and the providers and patients in the working interface. Likewise, the interdependence of all parts of the blueprint in a single system invalidates the claim of any individual in the system to be uninvolved in patient safety. The leader's role is to ensure the improvement of the organization with respect to all aspects of the blueprint, but most especially changing its culture—a responsibility that cannot be delegated. Changing culture starts with the leader and is a matter of sustaining climate change.

In the next chapter we will explore the nine dimensions of organizational culture—measurable dimensions that have been shown to influence and predict organizational safety most reliably.

CHAPTER THREE NINE DIMENSIONS OF ORGANIZATIONAL CULTURE

NINE DIMENSIONS OF ORGANIZATIONAL CULTURE

In the last chapter we defined culture as the shared, often unconscious values, attitudes, standards, and assumptions that govern behavior, especially in situations that lack written rules and procedures. Culture is "the way things are done around here." This definition, though descriptive, is insufficient when it comes to investigating culture's important role in organizational effectiveness.

Admittedly, organizational culture can be difficult to understand, because it is hidden. "The way things are done around here" connotes an indistinct collection of attributes. The values that drive the organization are frequently unstated and taken for granted. In one sense, everyone knows these things, but we lose sight of them. This slippery quality gives people the idea that culture is too vague a concept to be helpful.

In this book what we mean by culture is not at all vague. We mean a distinct, observable, and measurable set of attributes that we know to be predictably related to exposure to hazards in the working interface and thereby to the presence or absence of preventable adverse events.

Understanding your organization's culture in terms of the dimensions described here provides the insight needed to design targeted interventions that quickly build a strong safety climate and grow a culture in which safety is a driving value. While we do not claim to identify and understand *all* dimensions of culture, we do know those behavioral dimensions predictably related to safe and ethical behavior.

Measuring culture with the Organizational Culture Diagnostic Instrument

We said earlier that if a healthcare organization enjoys a strong cultural commitment to safety, the leader can delegate some of the tasks of improving systems—and that if the organization does not have a strong culture, these tasks should not be delegated. But what exactly does "strong culture" mean?

Since 1999, BST has examined more than 100 studies of variables that predict business outcomes, including safety. From this data we isolated nine critical dimensions (Table 3–1) that define organizational culture and have the greatest consequence for safety performance. We then devised and validated a diagnostic instrument—the Organizational Culture Diagnostic Instrument, or OCDI—with which to measure the distribution of the organization's strengths and weaknesses across the nine dimensions of culture. As you review the dimensions in the table, consider how each dimension manifests itself in your healthcare organization.

People in leadership positions frequently feel more positive about the organization's culture than do people working at lower levels in the organization. Leaders may be unaware of problems in the working interface that the people working there have just come to accept as the way things are done. For example, individual physicians may be quite comfortable and satisfied with the way they round on their hospitalized patients. But the nursing staff may feel excluded from important information and find too haphazard the process by which physicians gain the benefit of nursing observations. Nevertheless, the nurses may just accept this situation as "the way things are done around here." Or, as another example, physicians may be dissatisfied because of problems in the communication of their discharge orders but may have become resigned to the situation.

It is necessary to understand the organization's culture from the perspective of the people whose working lives are spent in the working interface because this venue is where the culture has its most immediate impact on patient and employee safety. Treatment

TABLE 3-1. NINE DIMENSIONS OF ORGANIZATIONAL CULTURE.

CULTURAL DIMENSION	DEFINITION
1. Procedural Justice	The extent to which the treatment team member[1] perceives fairness in interactions with his or her superiors[2] and in the superiors' decision-making processes.
2. Leader-Member Exchange	The relationship between the treatment team member and his or her superiors. In particular, this dimension measures the treatment team member's level of confidence that superiors will look out for his or her interests in addition to those of the organization.
3. Leadership Credibility	The perception on the part of the treatment team member that what leadership says is consistent with what leadership does; an explicit measure of the gap between what leadership says it will do and what it in fact does.
4. Perceived Organizational Support	The perception on the part of the team member that the organization cares about, values, and supports him or her.
5. Treatment Team Relations	The treatment team member's perception of his or her relationship with others on the treatment team. How well do they get along? To what degree do they treat each other with respect, listen to each other's ideas, help one another, and follow through on commitments made?
6. Teamwork	The extent to which the treatment team member perceives that working with team members is an effective way to get things done.

1 Included in the treatment team are all individuals who serve the patient directly or indirectly in the working interface, from registration assistant to nurse and physician to medical technologist examining slides in the laboratory.

2 By "superior" we mean anyone higher in the formal or informal organizational or professional hierarchy. In this sense, a unit nurse, for example, may have many superiors: a physician, the head nurse, a manager, a treatment team leader, and so on.

CULTURAL DIMENSION	DEFINITION
7. Safety Climate*	The extent to which the treatment team member perceives that the organization values the improvement of safety performance.
8. Upward Communication*	**The extent to which communication about safety flows freely upward. In part, the relative presence or absence of the tendency to shoot the messenger, i.e., to punish the bearer of bad news.**
9. Approaching Others*	The extent to which treatment team members feel free to speak to one another about safety concerns.
***Scales 7, 8, and 9 are specific to safety performance.**	

team members in the working interface are the first line of defense against exposures to hazard and thus a reduction in preventable adverse events. How the culture affects their safety-related actions is critical.

Understanding culture from the treatment team members' viewpoint can be achieved using a diagnostic instrument that measures their perceptions and compares them with normative scales from other organizations. These relative comparisons allow the safety leader to develop an intervention plan to make improvements.

Diagnostic instruments versus informational surveys

Diagnostic instruments and informational surveys are not the same (Table 3–2). A survey is a set of questions thought to be important. The organization has its workers answer these questions, it sums up the answers, and then it examines the data. Workers complain about "being surveyed to death." They complain that after going to all the trouble of taking the survey, they see few changes. As a result, the credibility of leadership suffers. Healthcare organizations need to do fewer surveys and take more action based on valid findings.

In contrast, a diagnostic instrument has an empirically validated history of predicting specific outcomes and employs a normative database so that one organization can be compared with another.[3] Leaders can pinpoint low-scoring areas and design improvements that will alter organizational culture and performance.

TABLE 3-2. DIAGNOSTIC INSTRUMENT VERSUS INFORMATIONAL SURVEY.

DIAGNOSTIC INSTRUMENT	INFORMATIONAL SURVEY
▪ Relies on established predictive relationships between survey scales and performance outcomes. ▪ Provides scores in relation to a database for comparison across organizations. ▪ Pinpoints specific areas for improvement.	▪ Employs a set of questions thought to be important to the organization. ▪ Provides a convenient way of identifying and summing up anecdotes currently passing among those in the working interface. ▪ May act out rather than reveal organizational bias in its very formulation because of the perspective it assumes, the questions it asks, and the terminology it uses.

The findings in the scientific literature upon which we constructed the OCDI have been confirmed in our consulting experience as well as in a series of studies we have conducted. Since 1999 we have used the nine critical dimensions of culture to help leaders lower their organizations' injury rates by improving their organizational cultures. Figure 3–1 summarizes our experience with the 94 organizations for which we had complete data at the time of the study. It shows 12-month concurrent injury rates by OCDI score. The important finding is that organizations scoring in the lowest third on the OCDI had the highest injury rates; those scoring in the

3 Certain other healthcare-related instruments fall between a true diagnostic instrument and an opinion survey. The Agency for Healthcare Research and Quality's (AHRQ's) Hospital Survey on Patient Safety Culture is an example. It measures some aspects of team functioning and safety climate; it does not measure the dimensions of organizational culture known to predict performance. AHRQ maintains a database of other hospitals that have used the survey as a benchmark, but the survey has not been shown to have predictive validity.

highest third had the lowest rates. The differences shown are all statistically significant. Figure 3–2 shows that the same relationship holds for organizations with low injury rates.

FIGURE 3-1. TWELVE-MONTH CONCURRENT INJURY RATES BY OCDI SCORE AT 94 CLIENT SITES.

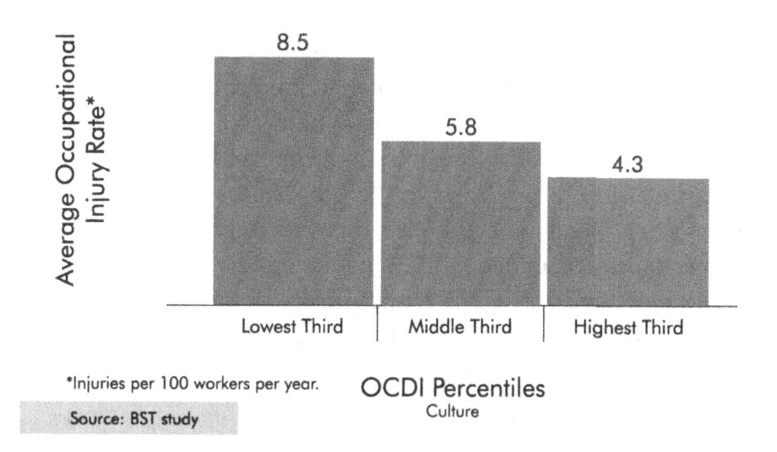

FIGURE 3-2. HIGHER OCDI SCORES PREDICT LOWER OCCUPATIONAL INJURY RATES FOR CLIENTS WITH OCCUPATIONAL INJURY RATES LESS THAN 3.0 (N=42).

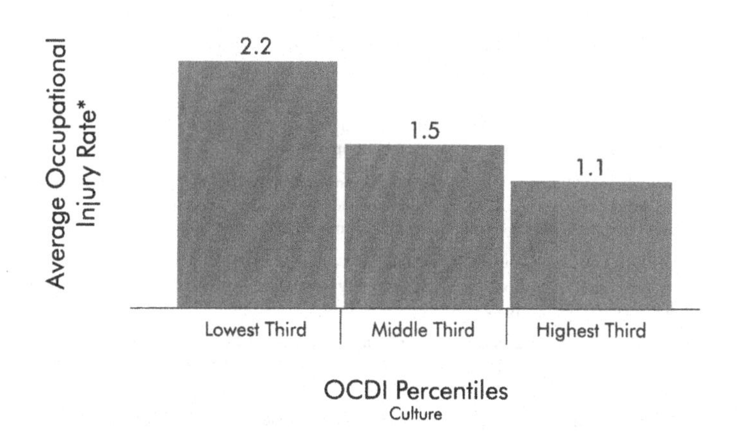

Clients in this study came from eight countries and 18 industries. The implications are straightforward: the wide variety of organizations, their different businesses, the variations in the strengths of their leadership, and the differences among their specific safety systems mean that higher OCDI scores, consistently across all scales, are more desirable than lower ones, no matter what the particular pattern of processes and results within the organization.

Each of the nine OCDI dimensions predicts safety outcomes (such as level of safe behavior, injury rates, and preventable adverse event reporting).[4] In addition to its patient safety[5] and nurse safety outcomes, safety climate has been shown to predict patient satisfaction, patient perceptions of nurse responsiveness, and nurse satisfaction.[6] OCDI scales also predict variables indirectly related to safety, e.g., turnover[7]; citizenship behavior[8–10]; job satisfaction; trust in the organization[11]; trust of employees, innovation, and creativity[12]; and organizational commitment.[13] So how the organization scores on the OCDI's nine dimensions also reveals how well the individuals in an organization function in relation to the organization's mission.

---

4 David A. Hofmann, *A Review of Recent Safety Literature and the Development of a Model for Behavior Safety* (Ojai, CA: Behavioral Science Technology, Inc., 1999).

5 Tal Katz-Navon, Eitan Naveh, and Zvi Stern, "Safety Climate in Healthcare Organizations: A Multidimensional Approach," *Academy of Management Journal*, 48 (2005): pp. 1075–1089.

6 Barbara Mark and David A. Hofmann, "An Investigation of the Relationship between Safety Climate and Medication Errors as well as Other Nurse and Patient Outcomes," *Personnel Psychology*, 59 (2006): pp. 847–869.

7 Gerald R. Ferris, "Role of Leadership in the Employee Withdrawal Process: A Constructive Replication," *Journal of Applied Psychology*, 70 (1985): pp. 777–781.

8 Jacqueline A-M. Coyle-Shapiro, and Neil Conway. "Exchange Relationships: An Examination of Psychological Contracts and Perceived Organizational Support," *Journal of Applied Psychology*, 90 (2005): pp. 774–781.

9 Mary A. Konovsky and S. Douglas Pugh, "Citizenship Behavior and Social Exchange," *Academy of Management Journal*, 37 (June 1994): pp. 656–669.

10 Jacqueline A-M. Coyle-Shapiro, Ian Kessler, and John Purcell, "Reciprocity or 'It's My Job': Exploring Organizationally Directed Citizenship Behavior in a National Health Service Setting," *Journal of Management Studies*, 41 (January 2004): pp. 85–106.

11 Jill R. Kickul, Lisa K. Gundry, and Margaret Posig, "Does Trust Matter? The Relationship between Equity Sensitivity and Perceived Organizational Justice," *Journal of Business Ethics*, 56 (February 2005): pp. 205–218.

12 Cynthia P. Ruppel and Susan J. Harrington, "The Relationship of Communication, Ethical Work Climate, and Trust to Commitment and Innovation," *Journal of Business Ethics*, 25 (2000): pp. 313–329.

13 Kerry D. Carson, Paula Phillips Carson, Ram Yallapragada, and C. William Roe, "Teamwork or Interdepartmental Cooperation: Which Is More Important in the Health Care Setting?" *Health Care Manager*, 19 (2001): pp. 39–46.

Cultural dimensions reflect organizational relationships

Treatment team members have four working relationships. They relate to:

- Other members of the treatment team
- Their superiors
- The organization as a whole
- Their patients in the working interface

For better or worse, the first three relationships measurably affect team member behaviors critical to safety in team members' fourth relationship—with patients.

The nine dimensions of culture map to the first three relationships by measuring treatment team members' perceptions of the relationships (Table 3–3). The focus of each dimension is either organizational or safety-specific.

Note that the first six of the nine cultural dimensions in Table 3–1 are not specific to safety. It may seem odd that safety outcomes are strongly influenced by cultural variables that appear to have little to do with safety. Organizations that achieve excellence in one area of performance, however, tend to achieve it in others. High-performing organizations tend to be good at many things. If the environment in an organization—which is what these six organizational dimensions address—is favorable, then safety initiatives will tend to be successful; if the environment is less favorable, the initiatives are less likely to succeed and to be sustainable. This fact has profound implications for the organizational value created by safety improvement.

Channels of influence

How do the nine critical dimensions of culture affect safety outcomes? Figure 3–3 displays the nine dimensions divided into three groups: organizational dimensions, team dimensions, and safety-specific dimensions. It also shows how these categories are related to each other and to safety outcomes. Each relationship shown is statistically significant, and thicker arrows denote stronger predictive relationships. For example,

TABLE 3-3. CULTURAL DIMENSIONS ORGANIZED BY RELATIONSHIP AND FOCUS.

	FOCUS OF THE DIMENSION	
	ORGANIZATIONAL	SAFETY-SPECIFIC
Relationship to superiors	▪ Fairness of decision-making processes (Procedural Justice) ▪ Strength of working relationships (Leader-Member Exchange) ▪ Honesty and consistency (Leadership Credibility)	Climate around bringing up safety concerns (Upward Communication)
Relationship to other members of the treatment team (and peers)	▪ Team effectiveness (Teamwork) ▪ How well group members get along (Treatment Team Relations)	Likelihood that treatment team members will speak to one another about safety (Approaching Others)
Relationship to the organization as a whole (and management and leadership as proxies for the organization)	▪ Organizational concern for treatment team members' needs and interests (Perceived Organizational Support)	Value of and priority placed on safety (Safety Climate)

note that the organizational dimensions strongly predict safety outcomes, both directly and indirectly (through their effect on team functioning and safety-specific factors). The causal paths depicted by arrows in Figure 3–3 have important consequences for understanding your organization's culture and designing effective interventions.

Starting with the arrows pointing to "safety outcomes" on the right side of Figure 3–3, note that each of the three cultural categories influences outcomes. Of these, the team dimensions have the weakest direct impact on outcomes. However, they, along with the organizational dimensions, strongly influence the safety-specific dimensions, which, in turn, have a strong, direct impact on safety outcomes.

FIGURE 3-3. RELATIONSHIP BETWEEN CULTURAL DIMENSIONS AND SAFETY OUTCOMES.

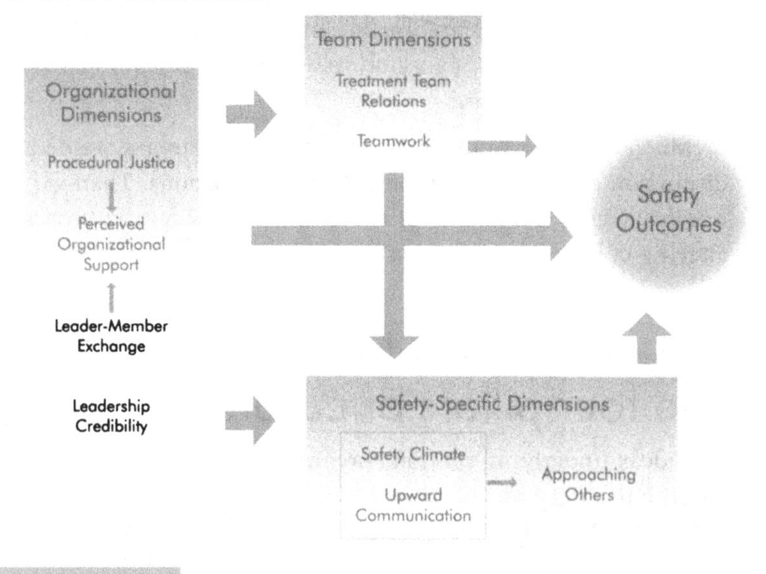

Source: BST analysis

The four organizational dimensions are the most powerful. As mentioned, this group of dimensions has a strong, direct effect on outcomes and a strong but indirect leverage on team functioning and on the dimensions that are related specifically to safety.

Note that the organizational and team dimensions are not safety-specific, which means:

1. For the safety-specific dimensions we could substitute equivalent dimensions related to other specific areas of organizational functioning, such as patient satisfaction, quality, or employee retention. We would likely find similar patterns of influence on these outcomes as we find for safety.

2. Sustained long-term improvement in safety is unlikely without attention to the more central team and organizational dimensions. An organization with poor relations between treatment team members and their superiors, or one with dysfunctional teams, will find it difficult to sustain gains that come from change efforts focused narrowly on safety. Long-term high performance in safety is much more likely when supported by a culture with strong organizational and team dimensions.

Much of the rest of this chapter is devoted to a more detailed discussion of each of the nine cultural dimensions. Then we will briefly discuss the impact of these dimensions on the process of altering culture change intentionally.

Organizational dimensions: The four pillars of culture

The ideas underlying the four organizational dimensions are not specific to safety but have to do with organizational functioning considered broadly. The organizational dimensions are:

- *Procedural justice:* Fairness of decision-making processes
- *Leader-member exchange:* Superior/treatment team member relationships
- *Leadership credibility:* Honesty, consistency, and competence
- *Perceived organizational support:* Organizational concern for team member needs and interests

These dimensions are the four pillars of culture because they reflect the level of goodwill and trust that underwrite the organization's capacity to function smoothly and efficiently. The customary definition of culture neglects the passion that lies not far beneath culture's surface. You can see in these four dimensions some of the sources of the passion. Justice, caring, and credibility easily arouse passion—especially when they become personal. The leader who sincerely desires to improve safety can tap into this passion.

Each of the four organizational dimensions predicts safety performance outcomes independently; together they are even more powerful. Of the four relationships that treatment team members have, the relationship with superiors is the most important to outcomes. To a large extent, a team member's superiors embody the organization. The quality of the relationship has a strong bearing on whether treatment team members believe the organization cares about their needs and interests (perceived organizational support). A team member who has a great relationship with the physicians and the head nurse, for example, also tends to give the organization high marks on perceived organizational support.

Procedural justice

Perceptions of the justice associated with interactions with superiors are powerful forces in an organization. Fair interactions are characterized by these qualities:

- *Consistency:* Decisions show predictability across persons and time.
- *Lack of bias:* Decision makers consistently steer clear of acting on personal self-interest.
- *Accuracy:* Decisions are based on good information and informed opinion.
- *Correctability:* Decisions made at various points of the process are open for appeal and reconsideration.
- *Representativeness (or "voice"):* The interaction reflects the basic concerns, values, and outlook of those affected.
- *Ethicality:* The interaction is compatible with the fundamental moral and ethical values of the organization and of those affected.

Treatment team members interpret as fair those interactions in which they and their patients are treated with dignity and respect. If leaders make decisions and interact in ways that are perceived as fair, team members assume they can perform their jobs without worrying about being at odds with the organization. How an organization handles concerns about employee and patient safety sends a strong message about whether superiors are playing fair. A high score for procedural justice indicates confidence that issues important to treatment team members will be addressed.

Leader-member exchange

The concept of leader-member exchange developed from attempts to understand exactly how leaders influence subordinates. Research shows that the *quality* of the relationship between the subordinate and his or her superior provides an important pathway mediating this influence. Specifically, leaders are more influential when workers trust that leaders are watching out for their interests in the workplace. This qualitative aspect of the relationship is the dimension measured by leader-member exchange.

One form of leadership influence is called transformational leadership (or sometimes just "leadership"). The transformational leader actively develops his people. When leadership has this quality, the treatment team member tries to achieve a goal not in anticipation of a reward or to avoid a punishment but because achieving the goal fulfills an organizational purpose important to both the leader and the team member. Transformational leadership exerts its effect principally through relationships and therefore represents an opportunity to build high levels of leader-member exchange.

Leadership credibility

Credibility can be viewed as an attitude held by one person toward another, based on the first person's observations of the other's behavior. Most research in this area has focused on the subordinate's trust in the leader and has focused on leader behaviors that lead to perceptions of trustworthiness. Various perspectives have been used to understand the development of trust. Research in social exchange theory reveals that leaders initiate the development of trust by acting in ways that provide benefits to followers (e.g., reducing uncertainty). Over time, the odds that followers trust the leader increase, and followers behave in ways that provide benefits to the leader (e.g., cooperation).

Perceptions of leader competence are a necessary but insufficient basis for developing trust. That is, team members are unlikely to trust a leader who is seen as incompetent, but competence alone does not ensure trustworthiness. Additional leader behaviors that build perceptions of trustworthiness include:

- *Consistency:* Demonstrates reliability over time and in various situations.
- *Integrity:* Exhibits consistency between word and deed, including telling the truth and keeping promises.
- *Sharing control:* Allows and encourages participation in decision making and delegation.
- *Communication:* Allows and encourages an open exchange of thoughts and ideas, provides accurate information, explains decisions, and provides timely feedback.
- *Demonstrated benevolence:* Shows concern for, consideration of, and sensitivity to subordinates' needs and interests; acts in a way that protects subordinates' interests; refrains from exploiting others for the leader's own benefit.

Perceived organizational support

Why would a treatment team member go the extra mile for the organization by, for example, actively participating on a safety committee? A team member's sense that the organization is concerned about his or her needs and interests strongly induces "extra mile" behavior. That is, workers who perceive they are supported by the organization are more likely to volunteer. This inclination also can be understood through social exchange theory. If team members believe the organization cares about and extends itself for them, they are more likely to extend themselves for the organization. It is also important that the organization's perceived support of treatment team members be seen as discretionary. That is, if a certain benefit or procedure is required by law or contract, workers generally will not see it as evidence of caring and concern.

Suppose a hospital strives for excellence in safety and goes well beyond what is required. Both the risk of personal injury and the exposure of patients to hazards represent emotional issues for treatment team members, and an organization that strives for excellence in safety thereby communicates its concern for both patients and members of the treatment team. That concern is likely to be reciprocated by discretionary effort on the part of team members.

Perceived organizational support is not the same as job satisfaction, though the two are often related. Workers who believe the organization cares about them are more likely to be satisfied. Perceived organizational support is an overall perception by workers of organizational commitment to them, whereas job satisfaction is an affective response (positive/negative) to specific aspects of the work situation (e.g., pay, physical working conditions, work schedules).

The strength of the relationship with superiors (leader-member exchange) affects treatment team member perceptions of perceived organizational support. It is as if team members see their relationships with superiors as representing the organization's concern for them. More generally, team members' perceptions of the extent to which the administration, managers, physicians, and (to a lesser extent) coworkers are trustworthy and supportive affect perceived organizational support as well.

We referred earlier to social exchange theory,[14] which is a useful way to think about why people in organizations do what they do. According to this theory, the important aspects of relationships (between individuals, or between an individual and a group) can be viewed as a series of exchanges or interactions in which the principle of reciprocity plays a central role. For example, if a nurse is treated with dignity and respect and feels supported by superiors, the likelihood increases that the nurse will reciprocate: his job performance, customer service behavior, extra-role behavior (going above and beyond), and loyalty tend to increase. On the other hand, the treatment team member who feels demeaned or disrespected is much less likely to engage fully in her work.

Table 3–4 shows the effects that strong or weak organizational dimensions can have in the working interface.

---

14 Jerald Greenberg and Russell Cropanzano, eds., *Advances in Organizational Justice* (Stanford, CA: Stanford University Press, 2001).

TABLE 3-4. MANIFESTATION OF ORGANIZATIONAL DIMENSIONS IN THE WORKING INTERFACE.

ORGANIZATIONAL DIMENSIONS:	▪ Procedural Justice ▪ Leader-Member Exchange ▪ Leadership Credibility ▪ Perceived Organizational Support
UNFAVORABLE PERCEPTIONS (Associated with Low OCDI Scores)	FAVORABLE PERCEPTIONS (Associated with High OCDI Scores)
▪ Taking hostile actions against coworkers ▪ Formalistic relations between superior and subordinate ▪ Few opportunities for subordinate input ▪ Low alignment between superior and subordinate goals ▪ Unwillingness by subordinates to go beyond formal job requirements ▪ Disengagement—low commitment to the organization or to its patient safety mission ▪ Low levels of initiative and out-of-the-box thinking ▪ Intentions to leave the organization ▪ Absenteeism	▪ Organizational citizenship behavior (going above and beyond the call of duty, such as volunteering for safety roles) ▪ Less resistance to change ▪ Empowerment of subordinates by superior and encouragement of subordinate initiative ▪ Mutual trust, respect, influence, and obligation between superior and subordinate ▪ Good two-way communication between superiors and subordinates and free exchange of information and knowledge within the organization and treatment team ▪ Treatment team functioning more cooperatively, efficiently, and effectively; higher levels of team member performance ▪ Positive perceptions of the value the organization places on safety ▪ Commitment to the organization and its patient safety mission ▪ Subordinate willingness to raise safety concerns and to seek help when needed ▪ Overall job satisfaction and satisfaction with superior ▪ Higher levels of organizational performance ▪ Less exposure to hazards in the working interface ▪ Fewer preventable adverse events

Team dimensions

The team dimensions include two aspects of treatment team functioning—how effectively the team gets work done (teamwork) and how well the team members get along (treatment team relations). Perceptions of these aspects are highly related but distinguishable. For instance, a team could be unproductive or ineffective but have members who get along well with each other. Alternatively, team members may not play nice with one another but may prove to be highly effective.

The two team dimensions are affected by perceptions of more fundamental issues in the organization (the organizational dimensions discussed earlier). An organization that has fair procedures, good relations between treatment team members and superiors, trustworthy leaders, and concern for team members tends to have well-functioning teams. Not surprisingly, how team members are treated sets the stage for team effectiveness and cohesion.

Team functioning also affects perceptions of the value the organization places on patient safety, the climate around raising patient safety issues, and the likelihood of team members talking to one another about safety-related behavior. These perceptions, in turn, affect safety outcomes, so teamwork and treatment team relations have both direct and indirect effects on patient safety outcomes (level of safe behavior, injuries, and preventable adverse event reporting).

Teamwork

Teamwork is affected by fair treatment of the members, by both superiors and peers; the level of trust in the team affects how well the team functions. In addition, many other variables affect treatment team functioning, including design of the work and the team (sociotechnical considerations), team composition, the general organizational context in which the team operates, and internal group processes. The teamwork dimension represents an overall assessment of these different forces on group cohesiveness and functioning.

TABLE 3-5. MANIFESTATION OF TEAM DIMENSIONS IN THE WORKING INTERFACE.

TEAM DIMENSIONS:	▪ Teamwork ▪ Treatment Team Relations
UNFAVORABLE PERCEPTIONS (Associated with Low OCDI Scores)	FAVORABLE PERCEPTIONS (Associated with High OCDI Scores)
▪ Hostile exchanges between team members ▪ Reluctance to take risks interpersonally ▪ Higher turnover ▪ Resistance to authority	▪ Higher team member satisfaction with other team members, the work, and superiors ▪ Greater likelihood of helping out coworkers ▪ Higher commitment to the team ▪ Higher team performance ▪ Higher levels of safety involvement ▪ Raising safety concerns with superiors ▪ Talking to one another about reducing working interface hazards ▪ Fewer working interface exposures and preventable adverse events

Treatment team relations

Social relationships within the treatment team affect important safety-related variables. When there are low levels of trust, team members are less willing to take interpersonal risks. On the other hand, if relations between group members are good, team members feel more comfortable interacting around safety issues and raising concerns.

In high-performing teams, members are more likely to identify with the team. Identification leads to trust among team members, which results in cooperation. Dysfunctional groups with a low sense of team identity have low levels of trust. Social relationships among group members strongly predict team member compliance with safety policies and procedures.

Treatment team relations are also affected by the leadership of the group. Supportive and trustworthy behavior by leaders tends to encourage trust among members of the team.

Table 3–5 shows the effects that strong or weak team dimensions can have in the working interface.

Safety-specific dimensions

The three dimensions specifically related to safety represent three different links between team members and safety outcomes such as preventable adverse events:

- Team members raise safety concerns (upward communication).
- Team members speak to one another about exposures to hazard (approaching others).
- Team members perceive that patient and employee safety is highly important to organizational leaders (safety climate).

The sense of responsibility is strongly influenced by perceptions of the safety climate, which also affect the dimensions of upward communication and approaching others. Relations with superiors and other team members as well as the sense of fair treatment by the organization and superiors all affect whether team members are likely to raise their concerns regarding patient safety.

Safety climate

The idea of a safety climate gained attention around 1980. Culture is a more fundamental concept. As we said earlier, safety climate refers to the prevailing, usually consciously held perception of leadership's priority for patient and employee safety at a particular time. It reveals to treatment team members with immediacy and urgency what leaders expect, support, and reward in a particular setting. While it makes sense to talk about the culture of an organization, it is less meaningful to speak of the climate of an organization; rather, it would be the organizational climate for a specific performance outcome, such as safety, reliability, cost, quality, or ethical behavior.

Specific measures of safety climate vary, but a common underlying theme is leadership commitment to safety. The underlying logic is that leadership commitment to safety is often manifested in visible support in the form of resources and programs. This support results in positive perceptions of organizational commitment, which affects how people perform in the working interface.

This positive chain of influence reduces hazards, exposures, and preventable adverse events.

Note that the administrative leadership often stands in the middle of these influences in that they own the task of evaluating and allocating resources. Resource allocation—especially financial—typically falls to the administrators who mediate between and among competing clinical services feeling the pinch of scarce means while trying to accomplish seemingly critical clinical ends. The administrative handling of both operating budgets and capital expenditure budgets often limits the depth of care available from a specific clinical service. For example, would the next million dollars in capital funds be better devoted to the initiation of radial keratotomies in the ophthalmology service or to expanding the availability of magnetic resonance imaging to a suburban intake facility? The trade-off is partly financial—a return on capital sort of analysis—and partly a matter of clinical strategy: who will this institution be to its patient base?

The administration often faces the task of brokering the answers to similar questions among multiple competing interests in the delivery organization. What gets attention in the consideration of the choices conveys a powerful consequence for the visible climate of the organization via its message about immediate priorities. Does the leader consider professional readiness to deliver safely on the promise of the new technology when making such decisions?

Safety climate is underwritten by the organizational dimensions. In particular, there is a strong relationship between perceived organizational support and safety climate. Team members who believe the organization cares about them also accept that leadership is committed to safety. Commitment to safety is one specific way in which organizational support can be demonstrated, and such commitment is a means by which a leader can make his or her ethical commitment to safety visible and tangible. This display of commitment represents an important opportunity for a leader who wants to build a strong safety climate.

Upward communication

Whether team members raise safety concerns to leaders also links organizational and team dimensions with safety outcomes. For instance, workgroups characterized by fairness and support have fewer patient safety mistakes and fewer employee injuries.[15] How do fairness and support result in better safety outcomes? One mechanism is that team members speak up about safety concerns. A superior who is fair and supportive is more likely to listen to concerns and to respond appropriately. Over time, the willingness of team members to identify opportunities for improvement and superiors' commitment to take action work together to reduce exposures to hazard and thus reduce injuries.

Team functioning (measured by the cultural dimensions of teamwork and treatment team relations) also affects the willingness of team members to raise safety issues with superiors. In a dysfunctional team, team members are more reluctant to bring up issues that might elicit negative reactions. Superiors who are open and responsive to upward communication about safety issues send a strong signal to team members that the organization values safety.

Approaching others

The upward communication dimension deals with team members' raising safety issues with superiors—often, these concerns consist of systems issues. In a healthy safety climate, treatment team members also speak up to one another about ways to reduce exposure (the upward communication dimension), even about the exposures they see each other creating. The more that team members are involved in hazard identification and willing to speak up about exposures, the safer the working interface becomes.

The cultural dimension of approaching others is related to both leader-member exchange and safety climate. If the leader is seen as valuing safety, the subordinate can reciprocate by speaking to others on the treatment team about safety.

---

15 Barbara Mark and David A. Hofmann, "An Investigation of the Relationship between Safety Climate and Medication Errors as Well as Other Nurse and Patient Outcomes," *Personnel Psychology*, 59 (2006): pp. 847–869.

TABLE 3-6. MANIFESTATION OF SAFETY-SPECIFIC DIMENSIONS IN THE WORKING INTERFACE.

SAFETY-SPECIFIC DIMENSIONS:	▪ Safety Climate ▪ Upward Communication ▪ Approaching Others
UNFAVORABLE PERCEPTIONS (Associated with Low OCDI Scores)	FAVORABLE PERCEPTIONS (Associated with High OCDI Scores)
▪ Team members are more likely to attribute the cause of an adverse event to situational elements even when team member behavior is a major factor.	▪ Higher levels of involvement and initiative ▪ Higher individual commitment to safety ▪ Stronger feelings of responsibility for the safety of fellow team members and patients ▪ Greater likelihood that team members will raise safety concerns with peers and superiors ▪ Higher levels of adverse event reporting ▪ Lower preventable adverse event rates

Team functioning also affects each individual team member's willingness to approach other team members about safety. In a high-functioning team with good interpersonal relationships, members are willing to speak up to one another, confident of getting a reasonable reaction. In contrast, reactions in dysfunctional groups are unpredictable—or predictably negative.

Table 3–6 shows the effects that strong or weak safety-specific dimensions can have in the working interface.

Exercise: Identify the cultural dimension

To become more fluent in the nine critical dimensions of organizational culture, consider an organization in which the following events have occurred. For each event, identify which of the nine cultural dimensions the event reflects and the effect it is likely to have on the organization's culture. (The nine dimensions are listed in Table 3–1.)

1. A surgeon uses abusive language in the operating room.
2. Out of fear of losing several high-revenue-producing doctors, the hospital administrator asks the patient safety representative to temper his demands.
3. A family physician yells at a nurse who frequently calls her at all hours for an interpretation of her orders.
4. A head nurse gives her friends preferential assignments and work schedules.
5. A charge nurse refuses to consider the personal needs of a particular nurse who asks for special consideration because his wife is sick.
6. Responding to patient survey data, the administration invests heavily in a new waiting room and other features designed to create a more spa-like atmosphere; it cuts most proposals for improved patient safety.
7. A nurse is fired for her part in an incident in which a patient died because the wrong fluid was accidently injected into his spinal canal.
8. Root cause analyses are performed but the recommendations are rarely implemented.
9. Physicians and managers frequently talk about the importance of safety, but few attend the meetings about safety. When they do attend, they seem to think that many of the issues are insignificant. They don't care for the proposed solutions, and they don't stay for the entire meeting because of other, more important professional obligations.
10. There are few upstream measures for patient safety, and the board of directors does not receive routine reports on safety outcomes.
11. There is a very low *reported* rate of adverse events but "everyone knows better."

12. Everyone thinks patient safety is important but few volunteer to serve on a committee working to improve it.
13. There is acrimony between pharmacy and nursing about where certain medications may be stored.
14. Recurrent miscommunications at discharge result in a high readmission rate.
15. People know the policy, but certain policies and procedures are routinely ignored and shortcuts taken.
16. Medications are often not administered on time.
17. At shift change, a bed in a busy emergency room staging area is found to be occupied by a homeless person who is enjoying breakfast, a newspaper, and a place out of the cold. He spent the night there, with everyone on the floor thinking he was someone else's patient.
18. A nurse believes that a physician has miswritten a medication order. She corrects it rather than talking to the doctor about it.
19. A physician sees the patient of another physician and comes to believe that the first physician has made serious diagnostic mistakes and has treated the patient incompetently. She does not discuss her concerns with anyone.
20. Physicians pay lip service to best practices but do not implement them.

Why do some organizations change more readily than others?

As pressure for improved organizational performance accelerates, team members are being asked to go beyond their traditional job duties and take more responsibility; e.g., to be a patient safety champion or to serve on a committee working on a systems issue to improve the safety of the working interface. In some organizations workers are easily engaged and rise to the challenge and even give discretionary time to ensure that goals are met. In others, change efforts meet resistance and workers are unwilling to extend themselves.

What determines the team members' responses to the need for change? Why do some organizations adapt easily and others struggle? Answering these questions adequately is important to improved healthcare safety, since many organizations will find it necessary to bring about fundamental culture change to reach safety excellence.

Culture normally functions to maintain standards and create continuity. Thus some resistance to change is natural. Nevertheless, organizations differ in the amount of resistance they encounter. The nine dimensions of organizational culture help reveal why some organizations change more easily and more successfully than others. Our research shows that organizations with high scores on the nine dimensions of organizational culture have higher levels of performance generally and better safety performance in particular.[16]

Change efforts do not occur in a vacuum. Organization members usually have long histories with each other. An individual

Organizations with high scores on the nine dimensions of organizational culture have higher levels of performance generally and better safety performance in particular.

nurse, for example, may have had thousands of interactions with specific physicians, administrators, managers, and peers. Some people are undoubtedly better at certain things and tend to focus on certain aspects of care more than others. Interactions such as these teach caregivers what is important to others on the team, how they are likely to be treated in various circumstances, and whether others are likely to do what they say they will do.

Perhaps surprisingly, the success of most change efforts depends more on perceptions of basic aspects of organizational life than on perceptions specific to the area to be changed. For

---

16 Thomas R. Krause, *Leading with Safety* (Hoboken, NJ: John Wiley & Sons, 2005), p. 64.

instance, improvements in patient safety at the nursing level often depend more on nurses' perceptions of how they are treated by their superiors than on perceptions of the importance of safety to the organization. This phenomenon means that, in order to facilitate change, leaders need to make especially sure their organization's culture is high functioning in its organizational dimensions—recurring procedural justice, consistent leader-member exchange, reliable leadership credibility, and a strong perception of organizational support.

Finally, improvement efforts are most successful when the organizational culture is well understood by leaders. It is then possible to build upon favorable dimensions and undertake targeted improvement where the culture is weak. It is also possible for the administrative leadership to measure dimensions of the culture in tangible terms and report the results over time as a means of communicating the tone at the top, in the middle, and at the bedside in order to fulfill the board's responsibility to monitor the culture and oversee the emergence of safety hazards and ethical risks.

Research into the organizational influences on safety outcomes has both confirmed some long-held beliefs and turned up some surprises. The link between a strong leadership commitment to safety and good outcomes is now well established. It is clear what a strong commitment means: being knowledgeable about safety in one's area of responsibility, assigning a high priority to safety in arenas such as leadership meetings, and taking action and committing resources to improve patient safety. Safety outcomes are often better in organizations in which superiors encourage subordinates to bring up safety concerns and in which treatment team members take the initiative to talk to one another about patient safety.

The surprises relate to aspects of organizations that don't directly concern safety. Safety results are better in organizations in which subordinates have good working relationships with superiors, in which superiors are fair in making decisions, and in which the team is both productive and amicable—in other words, organizations that function well and function fairly also stand out in safety.

Another surprise is that if treatment team members believe that leaders are trustworthy, honest, and consistent in general, team members exhibit higher levels of safety performance. Furthermore, if they believe the organization cares about their concerns and interests, they are more willing to extend themselves on behalf of the organization and its patient safety mission. If patient safety is valued in the organization, team members are prone to expend extra effort to work safely and to improve patient safety.

An organization's culture and safety climate are potentially among its pivotal resources, fostering adaptability, loyalty, retention, and safety. Or the culture can create a drag on leadership, making leading an unending uphill struggle. The leader leads in the context of the organization's existing culture and either enjoys the benefits or suffers the inertia the culture generates.

In the next chapter we examine leadership more closely, exploring how leaders create organizational culture and what a leader must do to build a culture in which leading is not a struggle. We also consider how improving safety performance becomes an institutional value.

CHAPTER FOUR QUALITIES OF A GREAT *SAFETY LEADER*

QUALITIES OF A GREAT SAFETY LEADER

Organizational culture has sometimes been described as an organism—as if it had a life and mind of its own. This is an important misconception. It may seem like culture grows and changes of its own accord, but we know the direct, causal connection between culture and leadership. "The way things are done around here" is largely determined by leaders and inherited or received by workers.

The leader is the crucial, active pilot who drives the development of culture. Culture is the residue of the leader's beliefs, thoughts, language, decisions, and actions. The organization's culture is the leader's responsibility. For example, suppose the most respected physician in your organization is a superb surgeon who does the seemingly impossible: he saves patients who others would find impossible to save. He thrives on challenge. He inspires others. They take up the challenge, no matter how difficult, and they, like him, give it their all. Every patient needing surgery wants this surgeon, and he attracts patients from far and wide. He is king. The organization is proud of him and what he stands for. How does this hero-worship affect patient safety?

Driving hard to save patients is obviously a good thing. The surgeon enjoys his hero status and is proud that others strive to participate in what he has achieved. He has no intention of jeopardizing patient safety. He would be horrified by the thought. Unfortunately, by the time this leader's attributes are translated into the culture, they may be distorted in ways he would not recognize, and may put patients at risk.

Heroism is not a team sport. It results from extraordinary individual effort. It takes considerable ego. It stems more from an exceptionally strong will than from communication, collaboration, and cooperation, and it may devalue these underpinnings of patient safety. People don't question, much less challenge, their hero. Furthermore, heroes may give run-of-the-mill cases short shrift, because heroism only shines under the bright light of daunting challenge. In such a culture the patient with routine needs may not receive the level of detailed attention required for safety's sake. And the surgeon himself may not set a good example of being personally engaged in patient safety because he feels that being a superb surgeon represents the most that should be asked of him. Thus, unintentionally, a culture may develop that puts patients at risk.

Or consider the new administrator who has successfully returned her hospital to financial stability by, among other things, reining in costs. Such an accomplishment is certainly laudable. Nevertheless, we must ask whether she considers the safety implications and messages the budget cuts send—even implicitly—about safety's value to the organization.

> Leaders often overlook how their behavior shapes the culture of their organizations.

Who the successful leader is—his or her values, beliefs, commitments, interests, choices, and ways of doing things—sends a message about what works and what success looks like "around here." The leader's attributes set an example and create standards for behavioral reinforcement, both positive and negative, within the culture.

Serving as an example for others and creating reinforcement criteria are two mechanisms by which "who the leader is" gets translated into "what the culture is." For example, when the administration allows the heroic surgeon even slight concessions of convenience that put patients at greater risk, other physicians may adopt the surgeon's apparently cavalier attitude toward patient safety and perhaps show little interest in serving on patient safety action teams. Or the chief pharmacist, following the

CEO's example, may rein in the pharmacy's costs by continually switching to the latest low-cost supplier, unaware of the potential confusion this creates in the working interface, where nurses are already stressed and finding it difficult to cope with ever-changing medication dosage forms, names, labels, and appearances.

None of this is to say that safety requires trading an intense commitment to doing the best for the patient in favor of exemplary fiscal performance. In fact, organizations with the healthiest organizational cultures are often those most committed to the patient's welfare in every way, and they are often most productive as well.[1,2] Nonetheless, leaders often overlook how their behavior shapes the culture of the organizations in which they work. More than one organization we have served has learned this the hard way.

Organizational leaders who want to be great safety leaders overtly ask, "How does who I am and what I do weigh on safety?" We begin to answer this question by looking at the critical aspects of safety leadership and how these aspects relate to each other and to the organizational outcome of a culture that values safety. When considered in the light of the nine dimensions of organizational culture (chapter 3), these aspects provide a broad, empirically based understanding of successful safety leadership and what this kind of leadership looks like when achieved.

The Safety Leadership Model

The Safety Leadership Model (Figure 4–1) is made up of four rings:

- Personal safety ethic (discussed next)
- Leadership style (discussed later in this chapter)
- Best practices (to be discussed in chapter 5)
- Organizational culture (covered earlier in chapter 3)

The three inner rings have to do with successful safety leadership. The outer ring represents the influence of effective leadership on the culture of the organization pivotal to business outcomes, including patient safety.

---

1 Barbara Mark and David A. Hofmann, "An Investigation of the Relationship between Safety Climate and Medication Errors as Well as Other Nurse and Patient Outcomes," *Personnel Psychology*, 59 (2006): pp. 847–869.

2 Michael Arndt, "How O'Neill Got Alcoa Shining," *Business Week*, February 5, 2001, p. 39.

FIGURE 4-1. SAFETY LEADERSHIP MODEL.

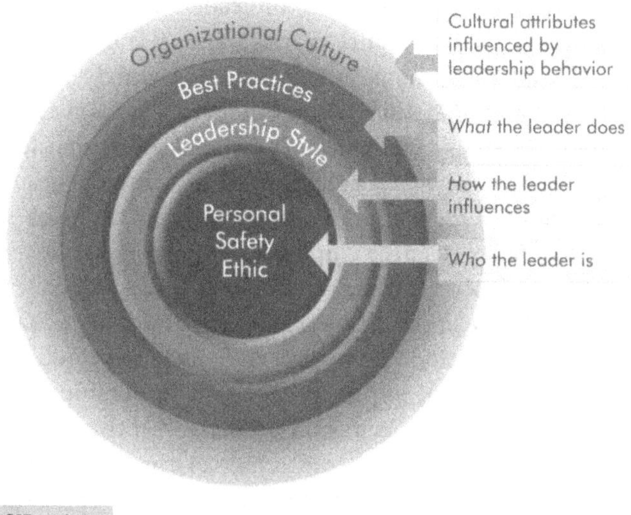

Source: BST analysis

The model can be read two ways: from the inside out, with the individual leader's personal safety ethic, leadership style, and practices emanating outward to the culture; or from the outside in, with the culture of the organization affecting the individual leader. Since our interest is primarily in how leaders influence culture, we will approach the model from the inside out. From the inside, each ring leads to the next and ultimately to business results. Who the leader is (her personality and values) sets the foundation for how she influences (her style), and what she does (her practices). Leadership practices shape the organization's culture, which, in turn, shapes safety results.

Who the leader is lies at the center of the model. The leader's personal safety ethic is at the core of safety leadership. No one can be an effective patient safety leader without genuinely valuing the safety of others. Personality, values, and emotional commitment are part of this ethic. Because our personalities are developed early in life and tend to change little in adulthood, a leader is more successful in achieving his desired ends when he understands how his personality affects his behavior and is experienced by other people.

How the leader influences—the leader's style—encircles his personal safety ethic. Over the years, researchers and scholars have described a plethora of leadership styles—laissez-faire, autocratic, charismatic, participative, transactional, theories X, Y, and Z—to name just a few. One widely researched style, transformational leadership, stands out as consistently predictive of business outcomes. The transformational leader focuses on the future, and her approach is strongly oriented toward developing her people. By going beyond her own self-interest, such a leader inspires employees to go beyond their mere short-term self-interest. The transactional leader, in contrast, focuses on current results and undertakes individual exchanges (of recognition, position, or money) to deliver expected results in the near term. We'll explore these two leadership styles, both of which have their merits and uses, later in this chapter.

The third ring from the center of the model addresses what the leader does—his behavior, or practices. The model positions practices next to style. For a leader to become a great *safety* leader, he must translate his core values, personality, and leadership style into safety-critical practices.

From working with safety leaders in hundreds of organizations, we have identified a set of seven best practices that are necessary to creating a vital safety climate and building an organizational culture that embodies safety as a value. We'll explain each best practice in chapter 5:

- Vision
- Credibility
- Action orientation
- Collaboration
- Communication
- Recognition and feedback
- Accountability

Organizational culture forms the fourth and outer ring. Culture is both the result of the personal safety ethic, leadership style, and best practices and the prerequisite for outstanding business success. As with the other rings of the leadership model, we let the empirical evidence drive our selection of the defining characteristics of organizational culture.

Measuring leadership with the Leadership Diagnostic Instrument (LDI)

The safety leadership attributes portrayed in the Safety Leadership Model can be measured with the LDI, a 360-degree survey instrument developed by BST. The LDI can be used to assess not just individuals but groups, such as the leadership team. Together, the LDI and OCDI (Organizational Culture Diagnostic Instrument) enable you to evaluate both your organization's leaders and the organization's culture with respect to safety. These instruments provide a clear picture of the current state and the desired state, thereby enabling you to plan your developmental journey and track your progress along the way.

> Each ring of the Safety Leadership Model predicts the subsequent ring.

When BST constructed the Safety Leadership Model and LDI, we required that all aspects be measurable and empirically validated. In 2006 we examined three years' worth of leadership data from the LDI along with concurrent site-level data regarding culture and safety performance. In a series of studies we found that direct reports' ratings on the LDI are valid and reliable measures of the model's underlying structure, and that each ring of the Safety Leadership Model predicts the subsequent ring, with culture being the strongest predictor of safety outcomes. That is, leaders' personality attributes and personal safety ethics predict their leadership style, which shapes their leadership behaviors, which prefigures the culture, thereby foretelling safety outcomes.

The numbers of sites and leaders participating in the overall study varied across individual studies depending on the availability of concurrent LDI, OCDI, and injury rate data for individual client locations. It is important to note, however, that this was the only selection criterion: each study used every site from which we had the required data from all three instruments. Sample sizes ranged from 39 to 94 sites.

We thoroughly examined the characteristics of the items and scale reliabilities. Reliability statistics were above 0.88 for every scale, which indicates that response consistency was high and the questions well understood, i.e., the scale reliably measures a single leadership characteristic.

One study (Figure 4–2) examined correlations between the transformational leadership style and the best practices. We found a significant positive correlation between subordinate ratings of each element of transformational leadership and each leadership best practice.[3] As we said earlier, leadership best practices consist of a set of specific *behaviors and activities* that safety leaders engage in effectively, while style is about *how* safety leaders perform these activities.

FIGURE 4-2. CORRELATION BETWEEN LEADERSHIP STYLE AND BEST PRACTICES.

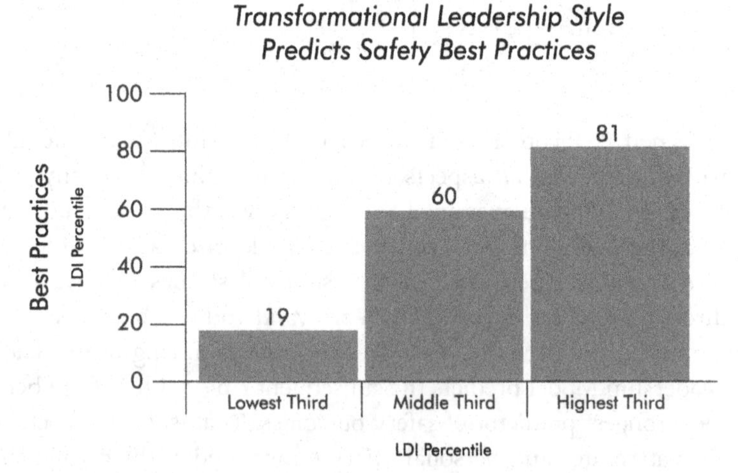

3 df (67), r =0.720, $P < 0.05$.

Another study (Figure 4–3) examined relationships between site-level top leadership teams' best practices and site-level culture. We found a significant correlation between subordinate ratings of each best practice and their ratings of each of the nine dimensions of organizational culture on the OCDI. In addition, we found that leadership overall (the aggregate of the seven best practices[4]) predicts culture overall.

FIGURE 4-3. RELATIONSHIP BETWEEN SITE-LEVEL TOP LEADERSHIP BEST PRACTICES AND SITE-LEVEL CULTURE.

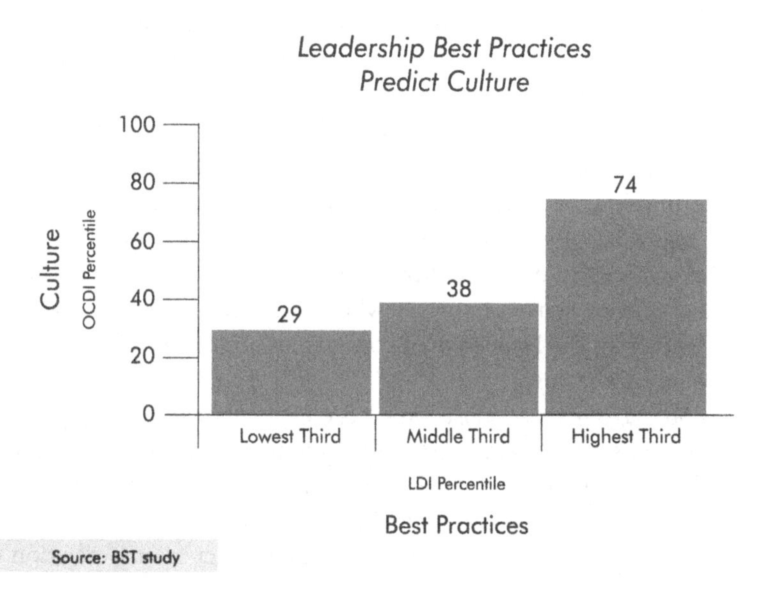

This series of BST studies, together with the study reported in the previous chapter demonstrating culture's effect on safety outcomes, shows that each ring of the model predicts its adjacent ring and safety outcomes. Because extensive data have been gathered on each ring and on the relationship between successive rings, individual leaders can comfortably trust the LDI as an assessment tool to compare themselves with other safety leaders and organizations worldwide and use this information to guide their personal safety leadership development.

4 df (69), $r=0.301$, $P<0.05$.

Personal safety ethic

At the core of the Safety Leadership Model is the leader's personal safety ethic. Culture receives and hosts the effects of a leader's values and behavioral standards—what he believes is important, what is acceptable, and what is not. Of course, what's important here is not what a leader *says* he values, but what he *actually* values—the ethics manifested in his personal behavior.

This means that you as the leader must "get it" before you start taking action to improve safety in your organization. You must be clear about your own personal safety ethic before you begin to lead safety. If you are not ready and start doing things anyway (even the right things), your actions may just complicate matters and even undermine your efforts. For example, if a leader gets on the patient safety bandwagon without thinking through the importance of each patient's safety, her actions run the risk of being perceived as empty—or worse, as hypocritical gestures. The workers in most organizations, both in and out of healthcare, have a keen sense of the lie when a leader's words and actions depart from the leader's own core values and beliefs. They are not fooled by the ruse.

So getting these most basic core elements right is the place to start. This tenet doesn't mean that you need to become the perfect safety leader before taking action. It just means that you need to understand how your personality drives your behavior, what your true personal values are, and how to act from a position of informed commitment.

A leader's personal safety ethic is a blend of the leader's personality, values, and emotional commitment to safety. Let's explore each one.

The leader's personality

Personality refers to individual differences in how people tend to think, feel, and act. We are all different, but what tendencies or dispositions make us different? Personality research addresses what is unique about the individual in terms of specific traits and attributes.

Psychological research in personality has been going on for more than 50 years. Hundreds of published studies have sought to define measurable personality traits. A wide variety of personality inventories and typologies have emerged, with varying degrees of success. While these approaches are somewhat useful in certain situations, a comprehensive grasp of personality traits and attributes eluded researchers for many years.

The Big Five

Personality research took a giant step forward when factor analysis (a sophisticated statistical technique) revealed that the dozens of personality attributes and traits that had been studied over the years could be reduced to five core attributes that define individual differences. Known as the Big Five (Table 4–1), the attributes have been subjected to extensive research:

- Emotional resilience
- Extroversion
- Learning orientation
- Collegiality
- Conscientiousness

TABLE 4-1. RELEVANCE OF THE BIG FIVE PERSONALITY ATTRIBUTES TO SAFETY LEADERSHIP.

ATTRIBUTE	CORE CONCEPT	SAMPLE DESCRIPTORS	RELEVANCE TO SAFETY LEADERSHIP
Emotional Resilience	How emotionally stable is the leader?	▪ How free is the leader from stress, anxiety, worry, frustration, irritability, anger, moodiness, and feeling overwhelmed? ▪ How easily are his or her feelings hurt?	▪ Much of a safety leader's work must be accomplished through his or her relationships. A degree of stability is required for successful relationships. This Big Five factor is associated with stable, successful relationships. ▪ A highly resilient leader will not be defeated by setbacks and difficulties, and others will likely experience him or her as strong and easy to be around. ▪ In the extreme, however, resilience can result in complacency, which will not serve the safety leader well.

TABLE 4-1. *(Continued).*

ATTRIBUTE	CORE CONCEPT	SAMPLE DESCRIPTORS	RELEVANCE TO SAFETY LEADERSHIP
Extroversion	How outgoing, talkative, positive, and dominant is the leader?	▪ How assertive? ▪ How easily does the leader mingle and make friends? ▪ How comfortable is he or she in taking charge and being the center of attention?	▪ Extroversion provides the basis for the relationships needed to foster good alliances devoted to safety improvement. ▪ In the extreme, however, it can produce expansive-but-empty visions and callous, self-defeating behaviors. ▪ An extroverted leader can be exciting and fun to work with—but also exhausting and domineering.
Learning Orientation	Does the leader prefer hands-on experience or a more academic approach to learning?	▪ How effectively and rapidly does the leader generate new and interesting ideas and try new approaches? How rich and difficult is the leader's vocabulary? ▪ Does the leader like to read or does he or she prefer to tinker? Is the leader's preference for book learning and theory—or for hands-on, practical experience?	▪ Learning orientation influences communication style. A learning preference shared with others can provide a platform for building enthusiasm about safety. ▪ But unless the leader can flexibly employ a style of communication appropriate to the learning preferences of those with whom he or she is interacting, the leader may inadvertently dampen the enthusiasm of the very people he or she is trying to stimulate.

ATTRIBUTE	CORE CONCEPT	SAMPLE DESCRIPTORS	RELEVANCE TO SAFETY LEADERSHIP
Collegiality	How interested in and sensitive is the leader to the needs and feelings of others?	▪ Does the leader have the capacity for gratitude, empathy, and consideration of others? ▪ Does the leader have the ability to relate to others and make them feel at ease? ▪ Does the leader have the ability to get along with, support, coach, and help others?	▪ Collegiality ensures that the leader has sufficient compassion for others to help them grow and to help them build. ▪ A collegial leader expresses empathy for patients and staff. It is easy to feel close and allied with a collegial leader. ▪ Collegiality may make others less objective toward and critical of the leader's ideas and behavior than they otherwise might be. ▪ At the extreme, collegiality may make the leader insufficiently demanding.
Conscien-tiousness	How im-portant to the leader is a struc-tured and reliable approach?	▪ Does the leader need regularity, a schedule, and punctuality? ▪ Does the leader attend to details? Is he or she well prepared? ▪ Does the leader have a plan and the ability to stick to it?	▪ Conscientiousness leads to effective attention to detail and an orderly process of implementing tasks that mitigate safety hazards for patients and employees. ▪ It is easy to depend on a conscientious leader because he or she has usually made a point to understand and master the details of the issue. ▪ At the extreme, conscientiousness can lead to rigidity and too much task focus at the expense of relationships.

The relevance of these five personality factors has been shown to pertain across leaders and staff generally, even across cultures, and to persist across time. The scientific literature shows that scores on these factors correlate strongly with specific aspects of leadership and career success.[5–7] One study gathered data on 244 families over 50 years and found that Big Five measures taken in childhood predicted the children's adult career success up to 50 years later.[8]

Measuring the Big Five

Behavior scientists have developed personality assessment instruments to measure the Big Five. (As part of our effort to help safety leaders become more effective, BST sometimes uses these instruments when we provide leadership coaching.) Research has shown that Big Five variables predict both leadership effectiveness and leadership emergence (who will emerge from the group to become the group's leader).[9] Extroversion and conscientiousness are the strongest predictors of leadership emergence. Not surprisingly, collegiality predicts success in jobs that require significant interpersonal interactions. Emotional resilience, extroversion, and learning orientation show statistically significant correlations with leadership effectiveness. Collegiality is also related to leadership effectiveness, but its correlation is not as strong.[10]

What bearing does this have on safety leadership in particular? Leadership effectiveness and leadership emergence are both highly relevant to the selection and development of safety leaders, since leaders who are effective in general are often the best candidates to serve as safety leaders. However, there are great safety leaders with many patterns of Big Five attributes. If you picture a great safety leader, you don't imagine someone who sits in his

5 Robert Hogan, Gordon J. Curphy, and Joyce Hogan, "What We Know About Leadership: Effectiveness and Personality," *American Psychologist*, 49 (June 1994): pp. 493–504..

6 G. M. Hurtz and J. J. Donovan, "Personality and Job Performance: The Big Five Revisited," *Journal of Applied Psychology*, 85 (2002): pp. 869–879.

7 Michael K. Mount, Murray R. Barrick, and Greg L. Stewart, "Five-Factor Model of Personality and Performance in Jobs Involving Interpersonal Interactions," *Human Performance*, 11 (1998): pp. 145–165.

8 T. A. Judge et al., "The Big Five Personality Traits, General Mental Ability, and Career Success across the Life Span," *Personnel Psychology*, 52 (1999): pp. 621–652.

9 T. A. Judge, J. E. Bono, R. Ilies, and M. W. Gerhardt, "Personality and Leadership: A Qualitative and Quantitative Review," *Journal of Applied Psychology*, 87 (2002): pp. 765–780.

10 Judge et al., "Big Five Personality Traits."

office all day and interacts only with his computer. Nevertheless, there are great safety leaders who do not score especially high on extroversion. They are able to perform successfully because they have learned to compensate behaviorally for their low extroversion. They can't change their personalities, but they can change how they behave and interact—their style and their practices—to compensate for those areas in which they score low and learn to moderate those areas in which they score high.

There is no doubt that personality structure is important to the leader's success. But if the personality you had as a child sways your leadership 50 years later, you are not limited or doomed. Low emotional resilience, for example, correlates with leadership difficulties and low career success, because low resilience can impair thought processes and therefore decision making. It may also result in relationship difficulties. Low resilience, however, does not *have* to result in these problems. A person with low emotional resilience can learn to compensate with self-management behaviors that others may never need to learn.

> We don't have to submit passively to the dictates of our personalities. We just need to expand our behavioral repertoire.

Using Big Five insights to improve safety

The very same personality attributes that represent our strengths can also become the source of our problems (Table 4–2). Armed with insight into our personalities and their relationship to our behavioral tendencies and effects on others, we don't have to submit passively to the dictates of personality. We just need to be willing to expand our behavioral repertoire. Knowing your Big Five profile will give you insight into the kind of leader you tend to be and where you need to compensate behaviorally to be more effective.

TABLE 4-2. BIG FIVE INSIGHTS FOR PATIENT SAFETY LEADERS.

BIG FIVE FACTOR	INSIGHTS
Emotional Resilience	High emotional resilience gives a safety leader a cool head in the midst of a safety crisis. On the other hand, the leader may appear aloof to others. With very high resilience, the safety leader may be impervious to feedback. Low resilience may result in strained interpersonal relationships and difficulty enlisting others in patient safety goals and objectives.
Extroversion	An extroverted safety leader is more likely to be out on the floor interacting with people in the working interface about safety. Engaging others in safety interactions is critical for safety leadership, and the extroverted leader finds this a natural thing to do. In the extreme, however, extroversion can be seen as superficiality. An introverted safety leader may not be easily accessible to others, and may have to find ways to compensate for his or her introversion.
Learning Orientation	A leader who is not oriented to learning may be uncomfortable having to learn and develop in the new areas that safety leadership requires. A safety leader who is extremely oriented toward learning may take action too late. The safety leader with an academic learning style will find books, articles, and lectures on safety interesting and stimulating. He or she will tend to approach others on the same basis, but this may or may not coincide with the other person's preferred style. Similarly, the safety leader whose style favors learning on the job will face analogous challenges when trying to train or inspire those with a more academic style.

BIG FIVE FACTOR	INSIGHTS
Collegiality	A highly collegial safety leader will tend to feel naturally the compassion needed for safety motivation. At the extreme, however, such a leader may not be sufficiently demanding. A safety leader with low collegiality will need to find ways to compensate, to devise other kinds of personal motivation, and to adopt new practices and behaviors designed to manifest his or her concern for the safety of patients and staff.
Conscientiousness	A highly conscientious safety leader will be inclined naturally to attend to the many details necessary to high-level safety performance. At the extreme, this leader may focus on tactical details to the neglect of important strategic issues or lose sight of the big picture. The leader may, for example, allocate too much attention to compliance issues and the letter of the law to the neglect of its spirit. At the opposite extreme, a safety leader with low conscientiousness may have grand ideas but little credibility.

Learning your profile provides guidance about what you should and should not try to change. For example, if you score low on extroversion, you would be fighting an uphill battle against your own nature to try to turn yourself into an outgoing, gregarious people person. Being the life of the party is not who you are. On the other hand, it might be relatively easy and critically important for you to spend a little additional time talking individually with each of your reports. Leaders scoring low on extroversion exert leadership influence through their relationships, just as extroverts do, but they may be more successful doing it one-on-one.

The CEO of a medium-sized hospital in the Midwest came to his current position through finance. His personality profile showed very high conscientiousness and high emotional resilience with relatively low collegiality. Consistent with this profile, interviews with his direct reports and the medical staff yielded perceptions of this leader as someone with very high standards who is detail oriented, demanding, unforgiving, and blunt—but also intelligent, high-achieving, and admired as well as feared.

Following a shocking sentinel event, the CEO adopted the urgent priority of significantly improving patient safety and appointed a team to design and implement the needed changes. Unfortunately, his impatience and intrusion into the workings of the team limited other people's willingness to engage. Ultimately, the project became mired in second-guessing, delays, and frustration. At this point this CEO sought the help of a consultant.

If you had been his coach, what would you have counseled?

The CEO and consultant identified several specific tasks the CEO could pursue and behaviors he could employ to rectify his problems with the team and get the project back on track. These behaviors and activities were essentially compensatory with respect to his personality structure. Note that the attributes for which he had to compensate were not weaknesses but strengths—attributes that had served him well up to this point in his career. He adopted these changes:

- He apologized to the team for his impatience and his excessively hands-on approach.
- He made his expectations explicit in the form of a team charter (as opposed to dropping in on team meetings and taking over to push for a solution to his latest concern).
- He stopped attending team meetings and instead instituted a review of the team's recommendations at specific milestones. He publicly committed to provide the resources for approved interventions.
- He received coaching on how to respond to the information he received from the team so as to be more encouraging and constructive.
- He became the spokesperson for the cause but let the team develop and implement the specific interventions.

Although this plan was in part a relief to him (since he no longer had to sit in on meetings in which he felt the team was spinning its wheels), he was not entirely comfortable with it. It seemed to him that he was giving up a great deal of control. He regained a sense of control by finding a number of venues at which could speak favorably about the changes the team was implementing and encourage others to participate. With coaching he became much more skillful at reining in his immediate impulse to dominate and criticize, and he learned to focus on and encourage even small movements in the right direction. This change in behavior paid off as more people came to support the project.

Knowing what you should compensate for behaviorally and understanding what you should *not* try to change are both significant benefits of having an individual assessment based on the Big Five. By shedding light on how your personality affects others, this kind of assessment also fosters your capacity for relationships. This is particularly important for safety leadership, which requires compassion—a quality not quite as important in other types of leadership. And compassion grows with the capacity for real and meaningful relationships, because relationships provide the ground for empathy. Empathy and compassion are necessary qualities for a great safety leader. Cultivating them does not mean giving up anything. On the contrary, it means gaining maturity and a more profound sense of humanity.

These considerations bring us to the next aspect of the Safety Leadership Model's center—the values and emotional commitment to safety that are part of the leader's personal safety ethic.

The leader's values and emotional commitment to safety

What motivates leaders to improve safety? Based on our experience with leaders ranging from first-line supervisors to CEOs, three primary motives drive safety improvement: compassion for others, a desire to unify the organizational culture, and a desire to earn profit sufficient to sustain the enterprise. Highly significant costs can be associated with patient and employee injuries. However, injury costs are more relevant to justifying the needed resources than to motivating leadership. Furthermore, improving safety to improve profits can undermine leadership credibility if others believe that profitability is the "real" reason leaders want to make safety improvements.

The predominant motive driving senior leaders to improve safety is compassion. This is true for safety leaders generally, whatever their place on the organization chart and their position on the treatment team, and whether the issue is employee or patient safety. The leader who works to improve safety is usually doing so out of a deep sense of integrity grounded in ethical principles—a belief that safety is the right thing to do. Compassion and motivation for excellence differ fundamentally from other business motives, including the drive for operating profits and personal success.

Compassion is the predominant motive behind senior leaders' drive to improve safety.

Organizations in which the individual's interests are not different from the organization's interests have many advantages, so leaders sometimes try to impose "values" in the hope of creating such alignment—e.g., by putting "Our Values" posters on the walls of a hospital or clinic. Such efforts often meet cynicism. However, it is difficult to be cynical about employee and patient safety, because being safe is intrinsically valuable. Highly effective leaders recognize that taking a leadership role in safety gives them an opportunity to create shared values in the organization. When values are shared, they have remarkable effects on organizational citizenship, on the ability of employees to work effectively as teams, and on overall organizational effectiveness.

Although "cultural unity" may be a secondary motive for building a culture that values safety, it can be critically important. Nothing undermines cultural unity faster than a seriously injured patient or a workplace that is perceived to be unsafe.

Safety as a personal value

The word *value* expresses the notion of worth or desirability. There are two categories of value: intrinsic and extrinsic. Intrinsic values have worth for their own sake; they are ends in themselves and have ethical import because they characterize what we think people should be pursuing. Extrinsic values, on the other hand, have worth only as a means to an end; their import lies in their utility. The worth of extrinsic values is derivative. They get their value because they advance intrinsic values.

Getting promoted, having personal power, and behaving ethically are all good examples of extrinsic values. Job titles, power, and ethics are all rightly regarded as valuable. They are useful because they enable a person to achieve things that command intrinsic worth. For example, money is valuable because it can further one's happiness and well-being. Ethical behavior is valuable because it can further not only one's own happiness and well-being but that of other individuals with whom one interacts. Happiness and well-being have intrinsic value.

A good leader is sensitive to the distinction between intrinsic and extrinsic values. This is how he or she keeps the organization focused and working to achieve its proper ends—whether that is performing surgery with technical excellence or scheduling to maximize the bottom line. In addition, however, a great safety leader is especially sensitive to intrinsic values. She believes in and is deeply committed to the worth of the individual—a belief that is deeply felt as an emotional commitment to the health and safety of each individual patient and employee.

A physician for whom patient safety issues have become especially important described his encounter with these emotion-laden, value issues in this way:

> Until I had to decide where to have my wife admitted for the workup and treatment of a potentially very serious illness, I had always just assumed that our hospital was good. But when I had to have her admitted, I had to consider whether our hospital would give her the best care. I must admit, it was a gut-wrenching decision and brought me face-to-face with whether we here at this hospital are practicing medicine with the care we should be. Could I trust my colleagues and the hospital to give her the kind of care she needed? It changed how I look at our hospital and whether we are doing right by our patients. It made me realize that it's real people we are dealing with, and that these things matter to them. Of course, I knew that before, but not in the same way I know it now. It's become personal to me, not just abstract.

Being an effective safety leader takes something over and above what it takes to be a good leader generally: it takes respect for the intrinsic value of others. This requires a significant degree of empathy, compassion, and maturity. These qualities are available to all leaders, but they must be cultivated and nurtured. Many leaders have not considered how to integrate these qualities effectively with their leadership roles, both medical and administrative, and with their personas.

A leader can express the compassion and emotional commitment required to be a great safety leader by not allowing himself to relate to patients and employees as mere resources or as paying customers, but rather by holding himself responsible to relate to them—in his interactions and even in his thoughts—as individuals.

A hospital CEO cannot know all the employees in her hospital individually. A busy physician may not have the time to get to know her patients personally. However, they both can muster sufficient empathy to be acutely aware of and respectful of patients' and employees' humanity: that every one of them is an individual who, like herself, experiences life as intrinsically valuable.

Why is it worth the effort? Because the patients and employees for whom we are responsible are intrinsically valuable, and this fact alone demands such effort on our part. The effort we make—or fail to make—shapes the values and culture of the organization. We cannot be effective in creating a culture that values safety unless we value and respect the intrinsic worth of the individuals in our charge. Our values play out in the culture. If a leader relentlessly and exclusively pursues extrinsic values (e.g., cost efficiency), or if that is how others perceive him, this sense begins to pervade an organizational culture in which motivation is directed away from safety toward these extrinsic values. Staff may then cut corners in the name of necessity, unconsciously neglecting patient safety.

Intrinsic values serve as a behavioral and cultural insurance policy for the benefit of the organization, the employees, and the patients. Intrinsic values create a barrier against doing the wrong thing. And because extrinsic values derive their worth from their relation to intrinsic values, a leader's neglect of intrinsic values ultimately disconnects staff behavior from its motivational source. People don't want to work just for the money. They want to work at something intrinsically valuable, something good. In our experience with organizational leaders, a leader who neglects intrinsic values—or worse, one who creates a conflict between extrinsic and intrinsic values—has a negative effect on employee morale and causes motivation to suffer. Conversely, where we have seen leaders call on employees' intrinsic values, greater engagement results.

So it is important for a leader to understand what he or she actually values. Compassion for others is a core feeling found in almost all of us. But the pressures and frustrations of a busy professional practice or day-to-day organizational life may drown it out. For some leaders, compassion comes easily; for others, it must be cultivated. In any case, it is the job of safety leaders to awaken and nurture compassion in themselves and others.

Patient safety versus employee safety

Leadership cannot credibly represent itself to treatment team members as promoting patient safety while neglecting employee safety. Yet drawing a distinction between these two facets of safety is exactly what many healthcare organizations do inadvertently. Often the employee safety and patient safety functions have different reporting relationships, command different budgeting priorities, and enjoy different status in the assignment of responsibilities.

From the cultural perspective, it is incoherent to claim to value safety but not all of safety. Such inconsistency undermines leadership credibility as well as leaders' efforts to improve the safety climate. Employee and patient safety move together. Creating a hazard-free working interface means a working interface that is safe for both patients and staff.

Moreover, a leader forfeits significant opportunities to engage team members in the safety effort by allowing differentiation between these facets of safety. Leadership's interest in employee safety provides a highly effective approach to building a culture that values safety. It also creates interest in and support for patient safety.

Valuing safety—part of the personal safety ethic at the center of the Safety Leadership Model (see Figure 4–1)—is necessary but not sufficient to create effective safety leadership. It is human nature that we often do not see ourselves as others see us. We may not have an accurate idea of what we value. We may want to be compassionate and respect the intrinsic worth of others. While our self-image may include these attributes, others might not judge us according to how we feel about ourselves, or even by our intentions. We are judged by our behaviors, the visible things we do and say, the decisions we make, and the way we communicate or fail to communicate.

This means that once we are clear about what we value and are in touch with the passion that underlies our values, we must translate this into behavior. Leadership behavior has two facets: *what* we do (leadership best practices) and *how* we do it (leadership style). We'll discuss style here and best practices in chapter 5.

Leadership style

The research literature has classified leadership style (the second ring in the Safety Leadership Model) in a number of ways. In recent years, the various dimensions and models have coalesced into two basic styles: transformational leadership and transactional leadership. (A third type, laissez-faire leadership, is also mentioned, but it amounts to an abdication of leadership responsibility and is thus not desirable to safety leadership.)[11] There is increasing evidence that transformational and transactional leadership are not mutually exclusive, but that different situations call for different styles. Great leaders are adept at using the mix that is appropriate to a given situation.[12]

Transactional leadership

Transactional leadership is based on the centrality of the exchange transaction—one might say, "the deal" between the leader and the worker. This style focuses on the connection between performance and rewards, and posits that people are motivated by self-interest. The word *transactional* refers to the quid pro quo nature of the relationship between the leader and her followers.

---

11 John Antonakis, Anna T. Cianciolo, and Robert J. Sternberg, eds., *The Nature of Leadership* (Thousand Oaks, CA: Sage Publications, 2004).

12 Bruce J. Avolio, *Full Leadership Development: Building the Vital Forces in Organizations* (Thousand Oaks, CA: Sage Publications, 1999).

A good transactional leader creates conditions that coordinate the leader's self-interest with those of her subordinates.

Transactional leadership can be active or passive. In the active form, the leader takes the initiative to communicate expectations and then monitors and reinforces performance. The research literature calls this constructive transactional leadership. In the passive version, the leader waits until something goes wrong and then responds with the appropriate consequence. This is called corrective transactional leadership or management by exception. The literature is very clear about the superiority of constructive to corrective transactional leadership. Unfortunately, few leaders avail themselves of the power of active transactional leadership, opting instead for the relatively weak, passive version.[13]

Transactional leadership, which is also called task-oriented leadership (and is regard by some as a form of management rather than leadership), is essentially conservative. It is an important leadership style for preserving existing cultural conditions and organizational practices and processes. It aims to get things done within the current context and works best in stable environments.

The transactional leader makes expectations and priorities very clear, actively monitors compliance, and reinforces successes. For example, the leader must ensure that the organization's safety-enabling and sustaining systems are in place and functioning well; he may need to do this personally by reviewing audit data about these systems and providing systematic feedback. If the leader is the CEO, he may need to ensure that his reports are on top of such issues. Or, to take another example, a physician in private practice needs to not only prescribe a medication but to follow up, monitor, and reinforce adherence to the treatment plan. These activities all call for a transactional style.

Transformational leadership

Transformational leadership is based on the development and nurturing of relationships. This style focuses on the future and is essentially developmental. It is most valuable when the task involves creating order out of chaos, breaking deadlocks, creating significant change in the organization, or developing the

13 Avolio, *Full Leadership Development*.

other person. It has also been called relationship-oriented, charismatic, or inspirational leadership. A transformational leader's role is to inspire others to go above and beyond their mere self-interest of the moment.

Several studies have shown that highly transformational leadership strongly predicts enhanced safety performance.[14] What connects this kind of leadership with improved performance? The connection is mediated by the aggressive, concrete actions that transformational leaders take to address identified safety concerns and issues.

Safety leaders are called upon to build a strong safety climate, and this inevitably involves changing the culture and developing both individuals and the organization. It may also require developing increased bench strength in safety leadership at various levels of the organization. The leader must create a vision of the strategic role that safety plays in the organization's future, challenge complacency, and develop others who can implement the culture changes needed to realize the vision. These activities lie at the heart of transformational leadership.

Because of the differences between the two styles of leadership, the virtues and behaviors that characterize them are different. Key differences are shown in Table 4–3.

Which style is more desirable for safety leadership? Research has shown that transformational leadership predicts a substantially higher level of performance than transactional leadership.[15] As you might guess by examining the differences between the two styles in Table 4–3, transformational leadership generates more energy and enthusiasm by mobilizing others' intrinsic values. For example, with transformational leadership the treatment team members are motivated by their feelings for the patient and their relationships with the leaders of the team rather than just working for their paycheck and their next day off. The key to this type of leadership is the effective appeal to others' intrinsic values.

---

14 Julian Barling, Catherine Loughlin, and E. Kevin Kelloway, "Development and Test of a Model Linking Safety-Specific Transformational Leadership and Occupational Safety," *Journal of Applied Psychology*, 87 (2002): pp. 488–496.

15 Barling et al., "Development and Test of a Model."

TABLE 4-3. TRANSACTIONAL LEADERSHIP VERSUS TRANSFORMATIONAL LEADERSHIP.

	TRANSACTIONAL LEADERSHIP	TRANSFORMATIONAL LEADERSHIP
Basis	Implicit contract	Personal relationships
Ethics	Is highly individualistic with an ethic of self-interest. Seeks mutual advantage through contractual relationships and fair play.	Is highly social with an ethic of the greatest good for the greatest number. Seeks to resolve value conflicts through mutually beneficial strategies and balancing the needs of all constituencies.
Motivation	Emphasizes the connection between performance and rewards; e.g., creates explicit expectations, monitors performance, and provides feedback.	Creates enthusiasm for the vision and loyalty to the leader, the organization, and the envisioned future; e.g., creates and passionately communicates a sense that all so-called preventable adverse events are actually preventable. Models and inspires.
Scope of work	Negotiates the performance/reward equation.	Expects people to go above and beyond their self-interest for the good of the group—including the patient—and exemplifies this in his or her own behavior.
Relationships	Task-focused, reliable and fair, but not necessarily personal. Insists on strong and effective following behaviors.	Personally involved with team members and patients based on their individual needs; e.g., makes an effort to help them achieve their aspirations. Over time, turns followers into leaders.

TABLE 4-3. *(Continued).*

	TRANSACTIONAL LEADERSHIP	TRANSFORMATIONAL LEADERSHIP
Emphasis	Emphasizes getting the job done. Does not encourage taking initiative or going outside the box. Views failures as impediments to production.	Encourages initiative and actively challenges old ways. Emphasizes finding new and better ways to do things. Views failures as learning opportunities.
Communication	Provides no more information than employees need to know to do what is expected of them.	Shares big-picture information widely and encourages others to communicate and express their opinions.
Effect	Big difference between the corrective and constructive types of transactional leaders, with the latter preferred by employees. The constructive type is generally more motivating, whereas the corrective type may be demotivating and foster resentment among employees.	Employees generally prefer the transformational style of leadership. Transformational leadership is associated with lower staff turnover and greater and more appropriate patient involvement in own care.

Research reveals that transformational leadership is especially more effective than a transactional approach when the leader has little or no control over how others will be rewarded for a satisfactory performance. This is often the situation in healthcare delivery, where treatment team members work without continuous direct supervision and are subject to many pressures that predispose to shortcuts. If a treatment team member does something in an unsafe way, the leader may only see that the job was done, not how safely it was done. The more removed the leader, the more acute this problem becomes. A physician can face a similar problem with her outpatients, whose motivations—such as a willingness to take their medicines as prescribed—she assumes (sometimes wrongly) to be aligned with her own.

Transformational leadership is also more effective when the leadership task at hand entails culture change. The transformational style is conducive to developing others to become leaders and motivating interest in new and better ways of doing things. It avoids the problem of blame and fosters cooperation on behalf of organizational goals.

Hence, we expect that a successful administrative leader in a large, complex healthcare institution—especially a leader who aspires to improve patient safety—frequently faces the need to bring transformational leadership skills to the fore. Moreover, this administrator may find transformational skills all the more necessary when dealing with the medical staff, whose professional stake—except in the extreme case of malpractice—may lie more in a clinical practice partnership than within the walls of the hospital.

The LDI measures not only the safety leader's best practices (discussed in the next chapter) but also the leader's style—the degree to which the leader aligns with these four characteristics of transformational leadership:

- *Challenge:* The leader provides subordinates with a flow of challenging new ideas aimed at stimulating them to rethink old ways of doing things. He challenges dysfunctional paradigms and promotes rationality and careful problem solving.
- *Engage:* The leader helps others commit to the desired direction. She coaches, mentors, provides feedback and personal attention as needed, and links the individual's needs to the organization's mission.
- *Inspire:* The leader sets high standards and communicates about objectives enthusiastically. He articulates a compelling vision and communicates confidence about achieving the vision.
- *Influence:* The leader builds a sense of mission-beyond-self-interest and a commitment to the vision. She gains the confidence, respect, and trust of others, considers the ethical consequences of her decisions, appeals to others' most important values and beliefs, and instills pride.

It is very important for safety leaders to develop their transformational leadership skills. But one cannot be an effective transformational leader without strong transactional skills. Both styles are necessary for great safety leadership. A natural disaster, for instance, calls for the forceful command and control skills of an efficient transactional leader to stabilize an inherently unstable situation and to get immediate and urgent results. However, such a situation also requires the supportive and developmental skills of the transformational leader to reassure and support disaster victims and build and sustain their hope and belief in the future. It behooves the leader to cultivate both leadership styles and to become sensitive about when to use each style.

Cultivating your leadership style

Unlike personality, style is something a leader can learn. Style is a matter of how one approaches opportunities, what one focuses on, ignores, chooses to emphasize, and delegates to others. These things are all behavioral and within the leader's control.

If a leader wants to improve his capacity for either transactional or transformational leadership, it is important that he first clearly understand his natural inclinations. Personality factors predispose to stylistic preferences but do not determine them. For example, extroversion predisposes to a transformational style. Leaders who do not score high on extroversion can nevertheless learn to be very strong transformational leaders. They may have to build on other facets of their personality, but they can be just as effective.

Using a 360-degree diagnostic instrument such as the LDI gives the leader a picture of where to strengthen his or her safety leadership. Doing so requires focusing on the appropriate leadership best practices—the subject of the next chapter.

CHAPTER FIVE

LEADERSHIP *BEST PRACTICES*

LEADERSHIP BEST PRACTICES

In the previous chapter we addressed the two inner rings of the Safety Leadership Model (see Figure 4–1): the leader's personal safety ethic (a combination of his or her personality, values, and emotional commitments) and leadership style (i.e., the extent to which the leader practices transformational leadership). Now we turn to the third ring in the model—the specific best practices that constitute safety leadership itself. How do great safety leaders behave? What are their virtues?

Leadership best practices are relatively stable dispositions of the leader to think, feel, and behave in certain sensible ways, which is essentially the notion of virtue as defined by the early Greek philosophers.[1] Back then, virtue did not have the connotation that it has for us today. The Greeks did not see virtues as burdens, duties, obligations, or constraints on the individual. Rather, they saw them as habits conducive to excellence, well-being, and happiness—what a well-lived life was all about. The virtues were desirable ways of thinking, feeling, and acting that anyone who reflected seriously on life would wish to develop. They were habits of character reinforced through reflection and by repeated action toward ends that were intrinsically valuable.[2]

We have all known, or known of, great leaders: people whose commitment, combined with excellence in leadership, has enormous positive influence. Bear in mind that health-care organizations need leaders at all levels, leading formally or informally. If we understand what effective leaders do in

1 Julia Annas, *The Morality of Happiness* (New York: Oxford University Press, 1993): pp. 48–52.

2 Ibid., pp. 27–46.

concrete, behavioral terms, we can help others like them develop throughout the organization.

As we saw in chapter 2 (see Figure 2–1), leadership is but one expression of healthcare safety excellence. Improving safety in an organization essentially means creating a strong organizational culture—and, as part of that, a strong safety climate (the more immediate perception of leadership's priorities with regard to safety). Within this environment, organizational sustaining systems and healthcare safety–enabling elements function at a high level as perceived by those who use these systems as well as by senior leadership. The working interface is continually improved through the reduction and elimination of exposures to hazard.

The most difficult aspect of safety improvement lies not in implementing safety systems and mechanisms—though all are essential to safety. The more difficult challenge is to create a culture in which safety is a driving value. Enter: leadership. Creating this kind of culture is something only leadership can do. To understand this claim, consider the difference between leading and managing.

Management is a tightly focused task orientation—seeing to it that things get done. Leadership has a big-picture orientation—determining the right things to do, plus *how* and *why* they are done.[3] Managing involves directing the tasks of other people via activities such as writing job descriptions, controlling the operating room schedule, developing a staffing plan, creating a budget, and so forth (Figure 5–1). Management engages people's minds and causes them to take action. Leadership, on the other hand, has more to do with the spirit with which actions are taken, how they are done, and why they are important. Leadership engages people's hearts and strengthens their connection to their work and to their own motivations.

Organizations that function best with regard to safety have both effective leaders and good managers—not just people who direct what is to happen next but also people who move the organization to understand why things need to get done, how they should be done, and why it all matters. Such understanding helps to build culture.

---

3 James M. Kouzes and Barry Z. Posner, *The Leadership Challenge* (San Francisco: Jossey-Bass, Inc., 1995).

FIGURE 5-1. LEADERSHIP VERSUS MANAGEMENT.

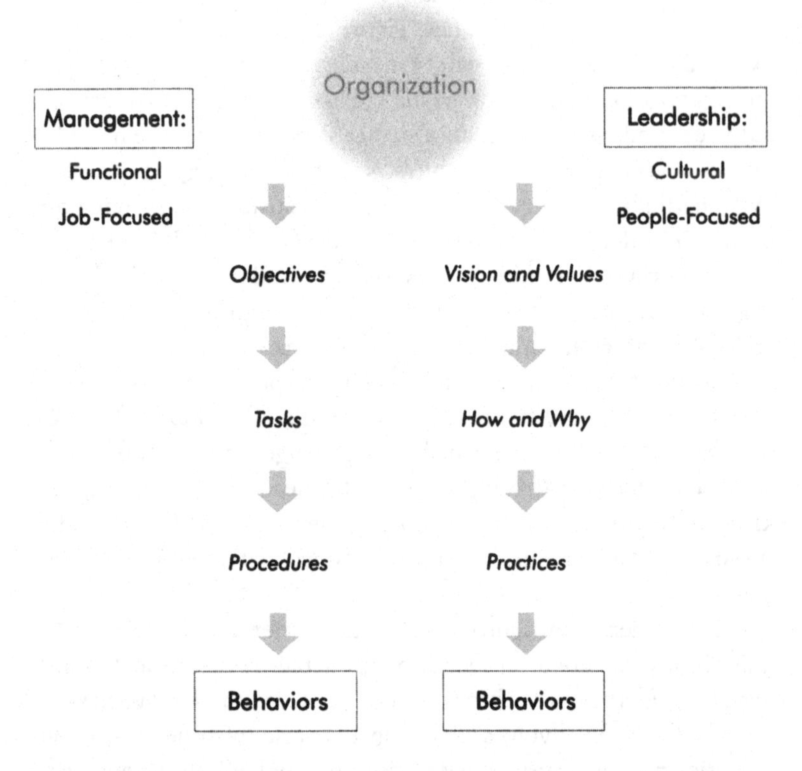

In working with safety leaders across many organizations, we have identified practices that all effective safety leaders do well. What these leadership best practices instill in the organizational culture is this: doing the right things for the right reasons and in the right spirit. The seven best practices for excellence in health-care safety leadership are:

- Vision
- Credibility
- Action orientation
- Collaboration
- Communication
- Recognition and feedback
- Accountability

Although these practices enhance leadership generally, we derived them with safety in mind. Effective leaders of all stripes are also effective safety leaders *if* they value safety, and effective safety leaders tend to be effective leaders in general. As we said earlier, these best practices develop in the leader through reflection and by acting on behalf of items that have intrinsic worth. In the case of patient and employee safety, the individual person commands intrinsic worth.

Thinking of best practices as a to-do list detracts from their value. Instead, think of them as challenges—opportunities to build character and develop the strengths you need to create a strong organizational culture. As a safety leader your job is to apply the best practices continually, consistently, and with the right spirit throughout your day-to-day business life. What is the right spirit? It means urgently ensuring the safety and well-being of employees, team members, and patients, because as human beings they are intrinsically valuable.

The seven practices are sequentially related and cumulatively developed; each builds on the prior practice. Safety leadership starts with the leader's *vision* and *credibility:*

> I have looked at our organization's culture and leadership data, understand our vulnerabilities, and have a clear picture of what our organization will be like when we remedy these vulnerabilities. I see the future state and can describe it in ways that others find compelling and actionable. I realize that my effectiveness depends in large part on how credible I am in others' eyes—otherwise, why should they take my vision seriously? Only if I have a record of consistency, a reputation for keeping my word, and a habit of voicing the truth—even when the truth is unpopular—will the things I say about our future have sufficient meaning to be influential.

Action follows vision and credibility:

> I have a vision and can talk about it effectively and credibly, but vision is not enough. I also act on my vision. When I see the facts clearly, my decisions follow naturally from them. I do not hesitate to take decisive action.

And the action includes others through *collaboration* and *communication:*

> I actively collaborate with others about safety issues. I consult with others before making final decisions that affect them. Since I know that collaboration entails communication, I foster excellence in communication about safety issues.

Finally, leadership requires being responsive to the actions of others through *recognition and feedback* and *accountability:*

> Having set the stage for safety excellence, I stand ready to provide feedback and recognition when I see behavior changes in the desired direction. When people change their behavior, I let them know that I notice and appreciate the change. I also know the value of accountability: I hold people accountable for the results that follow when the right things are done and when they are not done.

Vision and credibility form the foundation upon which the other leadership virtues are built. Both are primarily antecedent in behavior analysis terms: they trigger others to act. Action orientation, collaboration, and communication are also largely antecedent. The last two practices entail providing consequences. These practices follow the behaviors that are required for building a strong culture and reinforce them. (We will have more to say about behavior analysis and the importance of antecedents and consequences in the next chapter.)

Let's examine each best practice individually.

Vision

Many leaders either lack vision or have it but find it difficult to articulate. Others seem to be natural visionaries and take this ability for granted. A great safety leader holds a clear picture of the future state of safety in her healthcare organization and articulates that picture in a compelling way. She can "see" in some detail the desired future state and communicate this state to others. How does the desired future state differ from the way things are today? What kinds of things will people do and say that they don't now? How will decisions be made differently, and what assumptions underlie these decisions? If by some miracle we could instantly change the organizational culture and safety climate today, how would tomorrow be different?

A leader with vision:

- Behaves in a way that communicates high personal standards for safety
- Helps others question and rethink their assumptions about safety
- Communicates the organizational vision through word and action
- Demonstrates a willingness to consider and accept new ideas
- Helps people consider the impact of their actions on the safety of others and on the organization's culture
- Challenges and inspires people around the safety vision
- Describes a compelling picture of what the future could be

A clear vision is usually contained in a story that explains and demonstrates both the need for change and the prospects for a more desirable future. Such a vision mobilizes people. It sets expectations, clarifies ambiguities, and contributes a unifying direction to the organization by providing a common conception toward which people can work. By helping to enlist many people in the change effort, the vision brings greater flexibility to the culture and provides more "change experiments." This kind of culture, in turn, produces solutions that help the organization accomplish the desired future state.

Exercise: Vision

1. Reflect on the ways that your healthcare organization's present culture may undervalue the intrinsic worth of team members and patients.
2. Imagine three specific ways the culture would be different if things were perfect.

3. What would you see and hear that would indicate these differences from the current state?
4. How would you explain the importance of these differences to others?

The effective safety leader needs to be able to see the desired future state vividly and to describe it in concrete and compelling terms. What makes a vision compelling? In part, it's the ability of the leader to describe the vision plausibly and in detail, with enthusiasm and excitement. In part, it's the leader's personal credibility.

Credibility

Great safety leaders enjoy high credibility with direct reports, with other treatment team members, with patients, and with the larger organization. Deep knowledge and great skill, though important, do not ensure a leader's credibility. People believe what a credible leader says and trust him or her to tell the truth, even (or especially) if the truth is unpopular and unlikely to be well received. The credible leader is seen as free of personal agendas. Most important, the leader's actions and words are seen as consistent. Such a leader is transparent.

A leader with high credibility:

- Demonstrates personal concern for patients and team members
- Acknowledges his own limitations and errors
- Is believable
- Is transparent
- Is trustworthy
- Speaks inconvenient truths about safety
- Follows through on her commitments

Credibility creates respect for and loyalty to the safety leader. It makes leading easier because it makes following easier. A leader's credibility improves the quality of communication and builds a culture of integrity. In a culture that enjoys high credibility, people can rely on the commitments of others. This trait facilitates trust and cooperation.

Exercise: Credibility

1. Reflect on your commitment to patient and employee safety. Consider ways in which your credibility may be weak. For example, does your commitment to safety waver when it comes to difficult decisions or actions? Does your nonverbal communication make your commitment less believable than you would like?
2. Are you visibly uncomfortable when you need to challenge the thinking of a senior physician or administrator with whom you genuinely disagree?
3. What actions would help to remedy this tendency?

Your actions needn't be perfect. It is important, however, that your intentions be clear and that you persist. This brings us to the next leadership virtue, action orientation.

Action orientation

Great safety leaders eagerly take action on behalf of safety issues and actively seek opportunities to do so. This propensity reinforces credibility and tends to flow naturally from it. Safety issues arise and the leader makes decisions: How should we deal with physicians whose handwriting is illegible? How are we managing the trade-off between costs and safe staffing ratios? What is our plan for safely managing emergency room patient flow? Action orientation means that the leader is persistent and innovative and feels a sense of urgency about safety.

An action-oriented safety leader:

- Addresses issues proactively
- Seeks opportunities to make safety improvements
- Makes tough decisions with regard to safety
- Feels a sense of personal urgency about safety
- Is energetic about achieving excellence in safety

A leader's action orientation encourages a positive approach to change within the organization. Safety issues are avoided less often. The working interface becomes safer as exposures are mitigated or eliminated. Hazard mitigation creates a momentum for change, and improvements come more quickly and are more visible. Employees feel that leadership and the organization care about them and their patients, and this, in turn, improves leadership credibility and perceived organizational support. A positive cycle begins, and the culture begins to radiate optimism.

Exercise: Action orientation

1. Name three things you have an opportunity to take action on now that would improve patient safety.
2. Prioritize these actions. Which will you do first?

In all likelihood, to accomplish everything you want to accomplish, you will need the help of others. This requires collaboration.

Collaboration

Collaboration means working together, much as scientists and academics work together in intellectual pursuits. Collaboration also means soliciting and taking into account the views of others before making decisions. Collaboration is critical to effective safety leadership because safety success requires the willing involvement of people throughout the organization, at all levels. Creating a culture that supports safety requires that every employee and team member understand and embody the core concepts and related behaviors that constitute safety excellence. This engagement is likely to occur at a meaningful level only if the people involved feel they are important to its success. Participation and collaboration engender engagement; unilateral decision making shuts it down.

A safety leader who collaborates well:

- Inspires the willing involvement of others
- Engages others in safety decision making
- Works cooperatively with others to achieve safety excellence
- Solicits the views of others before making decisions

Collaboration builds a culture of communication, participation, and commitment. It makes solutions more workable, thereby making the change process less problematic. Collaboration fosters understanding and goodwill.

Exercise: Collaboration

1. Name three instances in which you usually make independent decisions that could be improved by collaborative decision making.
2. How will you need to approach others to win their agreement to collaborate with you?
3. How will the content and style of your communications with others need to change?

Communication

Leaders who communicate well and foster a culture of open communication enjoy the benefit of an informed workforce and improved decision making. Increased communication supports cultural alignment and error-free productivity. Good communication authenticates safety competence because the very nature of safety effectiveness requires that each team member understands what the safety issues are, how they are being addressed, and his or her own role in the patient safety solution.

A safety leader with good communication skills:

- Supports the top-to-bottom transparency of safety concerns and results
- Creates an atmosphere in which safety communication is safe, expected, and reinforced
- Shares safety-relevant information in a timely manner

Exercise: Communication

1. Name three situations in which patient safety would benefit from better communication.
2. What can and will you do to facilitate such communication?

3. How can you best recognize good communication?

Getting team members to understand their roles in patient and employee safety is not the same thing as inducing them to do what is necessary to achieve it. Stimulating new behaviors and then helping people to sustain them requires the skills of recognition and feedback as well as accountability.

Recognition and feedback

The core principle behind the best practice called recognition and feedback is that performance improves when leadership notices positive change and acknowledges it. The acknowledgment need not be formal or financial, but it should be consistent and genuine, especially when new behaviors begin to emerge. New behaviors require reinforcement in order for them to become an established part of the culture.

The great safety leader is tuned in to the behaviors of subordinates and other treatment team members, sets the expectation that safety-critical practices will be followed, monitors safety behaviors regularly, and provides soon, certain, and positive feedback (to be discussed in the next chapter) when the behaviors do occur. Negative feedback also has its place in certain situations. But in the great majority of cases, soon, certain, and positive feedback is a better way to sustain effective behavior.

Recognition and feedback do not refer to safety incentive schemes. We have written elsewhere about the inadvisability of these approaches.[4] Our experience is that safety incentive schemes usually have negative rather than positive effects on organizational culture.

A safety leader skilled in the use of recognition and feedback:

- Notices and acknowledges positive changes in safety activity levels and hazard mitigation accomplishments
- Gives added organizational visibility to internal best practices in safety

4 Thomas R. Krause and R. J. McCorquodale, "Transitioning Away from Safety Incentive Programs," *Professional Safety*, 41 (March 1996): pp. 32–36.

- Provides consistent, accurate, and timely safety recognition
- Provides positive coaching and guidance as needed

Consistency in recognition and feedback helps climate change become culture change. Leaders who provide an environment of clear expectations rich in accurate, immediate, positive, and corrective feedback are fashioning a climate of recognition and justice and cultivating the ground for a change in culture. The importance of this insight cannot be overestimated. We will discuss this at length in chapter 8, which explains how to foster culture change and create a strong safety climate for your organization.

Exercise: Recognition and feedback

1. Think of three circumstances in which you could provide feedback to others about their safety activities and performance.
2. For each circumstance, how will you give the feedback?
3. What could you say that would make your positive feedback less impersonal, more genuine, and more powerfully appreciative?
4. What could you say to make clear and replicable the specific actions you are acknowledging?
5. If you are acknowledging a failure, what can you say to be specific and actionable without adding personal blame to the feedback?

Accountability

Much has been written about accountability,[5] the last of the leadership best practices. In the safety area, *what* employees and team members are held accountable for matters greatly. Holding people accountable for incident frequency rates only makes sense if the numbers are statistically valid. More important, upstream activities that produce safety results should be measured and accountabilities established for them.

---

5 See, for example, Gerald A. Kraines, *Accountability Leadership: How to Strengthen Productivity through Sound Managerial Leadership* (Franklin Lakes, NJ: Career Press, 2001).

The effectiveness of accountability depends on having the other best practices in place and working well. We find, however, that in most organizations this safety leadership best practice gets first attention from leaders. The danger in this reversal is that naked accountability—accountability in a setting short on vision, action, recognition, communication, collaboration, and credibility, not to mention resources—breeds an environment of resentment and distrust. Employees are held accountable without being given the resources, information, leadership, support, and encouragement they need for success.

On the other hand, when leaders have command of the prior six practices and routinely employ them, accountability is easier to master and more efficacious. In this case, accountability reinforces expectations and priorities, communicates standards that let people know how they should act, and facilitates individual responsibility.

A safety leader who is skilled in accountability:

- Employs accountability in the context of the other six leadership best practices
- Creates an atmosphere of personal responsibility for safety
- Uses upstream safety accountability criteria

When used correctly in combination with the other leadership best practices, accountability builds culture.

Exercise: Accountability

1. For what specifically are people held accountable in your healthcare organization?
2. Are their safety accountabilities clearly defined?
3. Do their safety accountabilities conflict with other accountabilities?
4. Are the safety accountabilities expressed only as negatives (e.g., don't cause a medical error), or do the accountabilities define positive activities and behaviors as well?

Measuring leadership best practices with the LDI

How can a leader know where he or she stands in terms of these best practices? We have found that the most effective way to measure leadership best practices (and leadership style) is by systematically asking people around the leader about their perceptions of the leader. This makes sense because the growth of organizational culture depends on how others experience and internalize the leader's influence.

FIGURE 5-2. SAMPLE LEADERSHIP DIAGNOSTIC INSTRUMENT (LDI) REPORT WITH BEST PRACTICES SCORES EXPRESSED AS COMPARATIVE PERCENTILES.

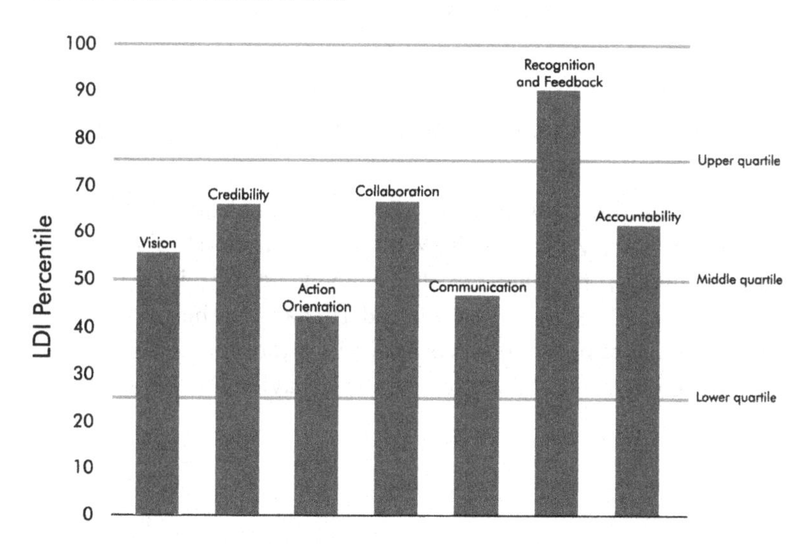

Various tools can be used for measuring leadership best practices. We use the Leadership Diagnostic Instrument (LDI),[6] which our clients have found to be a useful way for gathering both individual and group feedback. If the organization's intervention plan includes coaching to accelerate the acquisition of these leadership skills, the LDI can set the stage and provide valuable improvement targets.

6 As you will recall from chapter 4, the LDI is a validated and standardized diagnostic survey known to predict desired cultural outcomes and organizational safety performance.

On an LDI report, results are displayed as percentiles.[7] Thus each leader can compare his or her personal scores with a large database of other leaders' scores. This comparison leads naturally to the development of an action plan for the individual safety leader. Figure 5–2 shows a sample report.

Exercise: Changing behavior

If you were the leader whose best practices performance is reported in Figure 5–2:

1. What areas might you want to improve in order to strengthen your leadership capacity?
2. How would you go about it?
3. If the relative percentile indicates the need for change, which best practice(s) would you address and in what order?
4. Would you take the high score for recognition and feedback as an indication that no action is needed in this area?

Leaders build culture through their personal safety ethic (the core of the Safety Leadership Model), their leadership style, and their leadership practices. In this chapter we examined the best practices that outstanding safety leaders share. In the next chapter we introduce applied behavior analysis, a tool for systematically managing one's own behavior and the behavior of others.

7 Percentile is a rank ordering of the raw score data on a scale of 100 such that the 50th percentile is the middle of the distribution of raw score data.

CHAPTER SIX

CHANGING BEHAVIOR WITH *APPLIED BEHAVIOR ANALYSIS*

CHANGING BEHAVIOR WITH APPLIED BEHAVIOR ANALYSIS

Performance improvement inevitably requires behavior change. In the last chapter we discussed leadership best practices. When measured against these practices, leaders often strive to change their behavior so as to strengthen their leadership performance. But the need for behavior change does not apply only to leaders and to their practices. If the organization is to improve—at the level of the CEO, the physician, the supervisor, and the treatment team member—behaviors have to change everywhere. If the culture is to change, leadership behaviors that influence and support the culture need to change—but so do the behaviors of employees and treatment team members, who express by their actions the new cultural values. The leader's task, therefore, is to systematically and methodically identify and improve those safety-critical behaviors anywhere they occur, i.e., anywhere in the Blueprint for Healthcare Safety Excellence (see Figure 2–1). By vigorously embracing this task, the leader ensures sustained safety improvement.

Applied behavior analysis is a proven, scientific method for understanding and influencing behavior. An individual leader can apply this method informally or formally, and the leadership group can use it to build a mechanism[1] to drive change in behavior and culture.

It is relatively easy to determine how things should be done; the harder part is leading in a way that ensures things are actually

---

1 Recall what we mean by mechanism: a set of steps or system components that reliably lead to a defined result.

done that way. Most of us are all too familiar with the gap between the ways we would like to behave and how we actually do behave. We know we would like to lose weight, devote more time to our families, be more patient, follow our New Year's resolutions, exercise more, and get our important errands done. Nevertheless, we often fall short of doing what we intend. We remain frustrated and perplexed, sometimes by our own behaviors and often by the behaviors of others.

The same gap appears in organizations. The way we intend things to be done often differs from how they actually get done. This gap is so much a fact of organizational life that most of us have adapted to it and accept it as inevitable.

The way we intend things to be done often differs from how they actually get done. This gap is so much a fact of organizational life that most of us have adapted to it and accept it as inevitable.

The first step in closing any safety gap is to clearly specify the leadership-critical and safety-critical behaviors that define it. As indicated in the last chapter, the LDI can help with this task. So can a behavior analysis of incidents and close calls as well as focus groups aimed at identifying such gaps. Once we understand the behaviors that constitute the gap, we need to be clear about our objective. What do we mean by behavior change?

What is behavior change?

Choose any existing safety-critical procedure or practice—e.g., hand hygiene or the leadership best practice of vision—and then observe how the activity is actually done. You will usually find a difference between the intention and the actuality. Sometimes the activity is performed as you desire; at other times, it is not. If you look closely, you will see great performance variability.

When we talk about behavior change, we mean achieving a high level of reliable execution of the desired behavior or practice. We mean moving, for example, from 10% or even 85% desirable performance, to 99% or 100%. If a leader knows which behaviors are critical, this kind of behavior change is definable, measurable, and achievable.

Achieving high reliability challenges most of the organizations and leaders we serve. A behavior or practice that looks good on paper often does not translate into consistent action once leaders' expectations run up against the cold light of treatment team members' daily demands and priorities. This difficulty underscores the critical importance of understanding and influencing organizational behavior effectively.

Applied behavior analysis sheds light on the difficulty of behavior change and provides tools for making behavior change efforts successful. It provides a powerful methodology for understanding, measuring, and influencing behaviors of all kinds. You will find it an effective tool to help you increase your use of leadership best practices.

Applied behavior analysis is drawn from psychology. It is uniquely grounded in empirical research studies spanning more than 50 years.[2] Research in education, clinical psychology, and organizational improvement all demonstrate the effectiveness of applied behavior analysis as a tool for improving behavior.[3–5] Its power derives primarily from its concrete measurement of performance in observable terms.

---

2 B. L. Hopkins, "Applied Behavior Analysis and Statistical Process Control?" *Journal of Applied Behavior Analysis*, 28 (Fall 1995): pp. 379–386.

3 Jagdeep S. Chhokar and Jerry A. Wallin, "Improving Safety through Applied Behavior Analysis," *Journal of Safety Research*, 15 (1984): pp. 141–151.

4 J. L. Komaki, "Applied Behavior Analysis and Organizational Behavior: Reciprocal Influence of the Two Fields," In *Research in Organizational Behavior*, volume 8, eds. Barry M. Staw and L. Cummings (Greenwich, CT: JAI Press, 1986): pp. 297–334.

5 Thomas R. Kratochwill and Brian K. Martens, "Applied Behavior Analysis and School Psychology," *Journal of Applied Behavior Analysis*, 27 (Spring 1994): pp. 3–5.

Antecedents, behaviors, and consequences

Applied behavior analysis dissects a behavioral event into three elements:

- The target behavior
- Its antecedents
- Its consequences

A behavior is simply an observable act. An antecedent is an event or circumstance that precedes and triggers a behavior. A consequence includes any event or change that follows the behavior. For example, when the doorbell rings (antecedent), we answer it (behavior), and see who is at the door (consequence). Common sense tends to identify the antecedent (the ringing doorbell) as the cause of the behavior (answering the door). A central finding in applied behavior analysis research is that, although the antecedent is important, the consequence controls. Imagine a situation in which the doorbell rings repeatedly, but there is no one there when you answer the door. Perhaps the bell is malfunctioning, or maybe pranksters are ringing the bell and running away. In such

Although the antecedent is important, the consequence controls.

a case, the behavior of answering the door to see who is there is frustrated by the lack of the expected consequence. In fairly short order, you stop "automatically" answering the door. As soon as the ringing doorbell no longer reliably signals the presence of a visitor, it no longer elicits the behavior of approaching the door. By itself, the antecedent does not directly determine the behavior. Instead, antecedents elicit behaviors only because they signal or predict consequences.

In a nutshell, applied behavior analysis posits the following principles:

- Both antecedents and consequences influence behavior, but they do so differently.
- Consequences influence behavior powerfully and directly.
- Antecedents influence behavior indirectly, primarily by serving to predict consequences.

In addition to discovering that consequences are stronger than antecedents, behavioral science research has found that in the competition of consequences to control behavior, some consequences are stronger than others. Three qualities determine the potency of consequences:

- *Timing:* A consequence that follows immediately upon a behavior influences that behavior more effectively than a consequence that occurs later. Timing is scored as Sooner or Later, S or L.
- *Likelihood:* A consequence that is certain to follow a behavior influences that behavior more powerfully than an uncertain consequence. Likelihood is scored as Certain or Uncertain, C or U.
- *Significance:* All things being equal, a positive consequence influences behavior more powerfully than a negative consequence. Significance is scored as Positive or Negative, P or N.

Accordingly, *soon, certain,* and *positive* consequences have the greatest potency, or power to influence behavior.

A *reinforcer* is a consequence that increases the likelihood of a behavior's occurring in the future. *Reinforcement* means the use of supporting consequences to strengthen a behavior. Reinforcement thus provides a learning mechanism by which behaviors can be acquired. Behavior analysis goes to great lengths to specify the types of reinforcement, the conditions favorable to it, and the methods for selecting optimal forms of reinforcement and the timing with which reinforcement should be delivered in relation to the desired behavior.[6]

---

6 For a more detailed description, see *Behavior Analysis for Lasting Change* by Beth Sulzer-Azaroff and Roy G. Mayer (Wadsworth Publishing, 1991), or *Behavior Modification in Applied Settings*, 6th ed., by Alan E. Kazdin (Wadsworth Publishing, 2000).

ABC analysis

The above applied behavior analysis concepts are integrated and put to use in a tool called ABC analysis (ABC stands for antecedent-behavior-consequence), which helps reveal the causes of a behavior and influence its acquisition. Leaders use this tool to acquire best practice skills. ABC analysis has three steps:

1. *Analyze the undesired behavior.* Identify the target behavior you wish to understand or influence, and state it in the negative—the undesirable or problematic form of the behavior. For instance, if the behavior in question is action orientation, the undesired form would be a failure to act proactively. List the antecedents and consequences for the behavior, and list the potency of each consequence. This analysis gives you insight into why the problematic form of the behavior occurs and provides the data you need to develop a change strategy.
2. *Analyze the desired behavior.* State the target behavior in its desirable, nonproblematic form. List its antecedents and consequences, plus the consequence potencies.
3. *Create an action plan.* Write down the steps you will take to ensure that you have the right antecedents and consequences to acquire the desired behavior.

Let's say you are a leader who would like to strengthen your capacity for action orientation. As we recall from chapter 5, an action-oriented leader addresses issues proactively with urgency and persistence, seeks opportunities to make safety improvements, and makes tough decisions in order to improve safety.

Step 1: Analyze the undesired behavior

To help you understand why people see your level of action orientation the way they do, you might begin by gathering more information—e.g., by discussing your LDI score for this best practice with those who scored you. In this way, you get an idea of what they had in mind. Alternatively, you might ask your coach to interview your direct reports to gather this information. Once you have a clear picture of the issues that are relevant to your action orientation score on the LDI, you complete a step 1 chart, which might look something like the sample in Table 6–1.

Although the chart displays the ABC columns chronologically—antecedents, behavior, consequences—you fill in column B first. You describe the undesired version of the behavior and then fill in columns A and C. Do this analysis from your own perspective: the antecedents and consequences should be what really trigger and reinforce your enacting the undesired version of the behavior.

Although it falls outside the scope of behavior analysis as a discipline to do so, it may be useful to consider as well your internal perceptions. For many people, what they *feel* is motivating.

For example, although it is factually certain that if you fail to take action on a patient safety issue, eventually a patient will be injured, it doesn't *feel* certain. It *feels* uncertain because we don't know when our inaction will catch up with us—perhaps never. To see the power of emotion in the decision not to act, consider how you would behave if you actually felt certain that your inaction today would cause a patient to be injured later today. You would be highly motivated to act.

As another example of the importance of the personal perspective, consider someone who is addicted to smoking. He knows—in an impersonal, conceptual way—that it is highly likely that cigarettes will damage his health, but that often is not how he feels. He would be less likely to light up today if he really felt that doing so would cause a malignant mass to show up on his chest x-ray tomorrow. Instead, he may occasionally feel afraid that he might develop cancer some day, but he reassures himself with the fact that he feels fine now.

When you fill out column A you are answering the question *why,* explaining to yourself all the reasons for this undesired behavior that come to mind. In a real consulting engagement, this charting would be private, shared only with a coach (if you so chose), so be honest about the antecedents.

In column C you fill in all the consequences, good and bad, of the behavior. For instance, eating a large piece of chocolate cake is both pleasurable and problematic—it tastes great, but you may worry about your waistline the next day. Explore the consequences of the behavior to both you and others.

TABLE 6-1. ABC ANALYSIS, STEP 1—ANALYZE UNDESIRED BEHAVIOR.

A (Antecedents)	B (Undesired Behavior)	C (Consequences)	CONSEQUENCE POTENCY		
			Timing	Likelihood	Significance
My study of the problem has not been completed.[7]	Failure to take action proactively or on a timely basis on patient safety issues.	By not dealing with this issue I have more time for other things.	S	C	P
I am busy with other things.[8]		By putting it out of my mind, I experience reduced anxiety.	S	C	P
I don't have the data I need to make the decision.		Important safety problems don't get resolved.	L	U	N
I don't really understand the field of patient safety.		I remain unaware of critical problems.	S	C	N
I am not entirely comfortable with the issues in this area.		Hazards are not removed from the working interface.	L	U	N
		Patients are injured.	L	U	N
		I am caught off-guard when an incident occurs.	L	U	N
		I get other things done.	S	C	P
		I get relief from worrying about the complexity and work involved in patient safety issues.	S	C	P

Note: S or L = Sooner or Later (timing); C or U = Certain or Uncertain (likelihood); P or N = Positive or Negative (significance). The most potent consequences are S/C/P—soon, certain, and positive.

7 As you recall from chapter 4, the Big Five are the elements of personality that distinguish one person from another. This antecedent may be related to a high Big Five score on conscientiousness.

8 This antecedent may be related to a low score for the leadership best practice of collaboration, the tendency to solve problems oneself rather than involve others in their solution.

TABLE 6-2. ABC ANALYSIS, STEP 2–ANALYZE DESIRED BEHAVIOR.

A (Antecedents)	B (Desired Behavior)	C (Consequences)	CONSEQUENCE POTENCY		
			Timing	Likelihood	Significance
Adverse events occur, and I recognize that much needs to be done.	Taking action about patient safety issues proactively and promptly.	Fewer hazards in the working interface.	L	C	P
Root cause analysis reveals that systems are lacking.		Fewer patient safety incidents.	L	C	P
My boss tells me we have to improve patient safety.		I feel more in control of my time and of this area of my responsibility.	S	C	P
I receive a performance report on patient safety outcomes.		I am perceived and acknowledged as a more effective leader.	S	C	P
I read a book on patient safety.					

Note: S or L = Sooner or Later (timing); C or U = Certain or Uncertain (likelihood); P or N = Positive or Negative (significance). The most potent consequences are S/C/P—soon, certain, and positive.

In the far-right column, code the consequence's potency. Does the consequence occur sooner or later? Is its likelihood certain or uncertain? Positive or negative?

Step 2: Analyze the desired behavior

Write down the desired version of the behavior. Then identify its antecedents and consequences. Your analysis might resemble Table 6–2. Again, this leader started with column B and then filled in the remaining columns.

Step 3: Create an action plan

Use the information you generated in step 2 to develop an action plan. The plan should provide both antecedents and consequences that will help you acquire the desired version of the behavior and deal with potentially problematic antecedents and consequences (Table 6–3).

TABLE 6-3. ABC ANALYSIS, STEP 3—CREATE ACTION PLAN.

DESIRED BEHAVIOR: Spending adequate time on patient safety activities.

ACTION	BY WHEN
Find specific people to provide feedback for specific targeted behaviors.	
Attend one-day Patient Safety Academy to get the training I need.	
Add standing items to the agenda to discuss patient safety in my key meetings; elicit input on the issues.	
Request data from the safety and quality officer that I can use proactively.	
Make an agreement with my boss that she will put patient safety in my goals statement and provide me with positive feedback for achieved results.	
In six months, repeat subset of the LDI to see whether I've strengthened my action orientation score.	

Putting the tools to work in your organization

Applied behavior analysis and ABC analysis represent a uniquely effective tool for improving behavioral execution throughout an organization. Leaders can apply this powerful tool not only for their personal development as safety leaders but also for organizational development. To do so, they think systematically and in behavioral terms about the changes they wish to see in others and use ABC analysis to develop action plans to support these changes. Then they follow up with soon, certain, and positive consequences to reinforce changes in the desired direction. The most effective way to apply this methodology is to build mechanisms into your organization to provide positive reinforcement—recognition and feedback—for performing safety-critical behaviors at a high level of reliability.

Our client experience teaches us that well-intentioned change initiatives most often fail because they do not systematically apply applied behavior analysis principles. The greatest failure lies in relying too much on antecedents and giving insufficient attention to the consequences needed to back up the antecedents. Policies, procedures, recommendations, and training—all antecedent interventions—are usually the first things leaders think of when they want change. Remember: antecedents are only powerful to the extent they predict consequences. Unless the organization attends to consequences, antecedents are no better than the futile, archetypal antecedent strategy: train-and-hope!

When we talk about attending to consequences, we do not mean punishment. Both positive and negative consequences influence behavior. Which is more powerful depends on several considerations. Overall, positive consequences foster organizational change more effectively. Their use fosters communication and cooperation. Positive consequences also create an optimistic atmosphere. Behaviors changed by positive consequences tend to generalize more readily, and positive consequences lack the undesirable side effects on culture often associated with negative consequences.

What kinds of consequences can an organization provide to encourage the behavior needed for behavior and culture change? Clearly, the consequences have to be consequential to the person receiving them, but beyond this, the answer to this question is controversial. Some say that to be effective, consequences have to be financial—and the stronger the better. Our experience leads to a contradictory conclusion. Safety incentives, e.g., tangible rewards contingent on incident rates, often prove counterproductive. We have written extensively about this conclusion elsewhere.[9–10] So here we will just say that we do not recommend their use.

Attending to consequences does not mean punishment.

To the contrary, we commonly see dramatic and documented improvements in safety with no material incentives whatsoever. If a company or organization is committed to the broad application of incentives, we urge attaching them to upstream behaviors and leading indicators, not to downstream incident rates. The latter often drive underground the vital reporting of near misses that allow the mitigation of risk and the avoidance of hazards. For safety improvements and hazard mitigation, our experience confirms that it is most effective to employ positive feedback as the primary soon/certain/positive consequence.

In this chapter we explored how a leader can use applied behavior analysis and ABC analysis to strengthen his own capacity to exercise these best practices, and we have touched on their use in organizational change. Small gains in these areas turn into large gains in culture improvement and safety performance. Best practices, applied behavior analysis, and ABC analysis have the virtue of focusing on a phenomenon that is tangible and easily measurable: people's behavior. These tools enable the leader to acquire and strengthen desirable, safety-critical behaviors.

---

9 Thomas R. Krause, "Motivating Employees for Safety Success," *Professional Safety*, 45 (March 1999): pp. 22–25.

10 Thomas R. Krause, "Moving to the Second Generation in Behavior-Based Safety," *Professional Safety*, 46 (May 2001): pp. 27–32.

But even as leaders use these tools, they can make mistakes of judgment about exactly when and how to employ them. Such mistakes are not limited to the employment of tools; they intrude into administrative and clinical decisions as well, where they may lead directly to an increased level of hazard in the working interface and to medical errors. In the next chapter we will explore how cognitive biases can give rise to these errors.

CHAPTER SEVEN

PROTECTING YOUR *DECISION MAKING* FROM COGNITIVE BIAS

PROTECTING YOUR DECISION MAKING FROM COGNITIVE BIAS

Cognitive bias refers to systematic distortions of decision making that are common to all human beings. These distortions arise from the inbuilt heuristics we use to simplify the large quantities of data with which our nervous systems routinely deal. Some researchers also include emotional prejudice as a mechanism underlying some of these biases. Whatever their underlying causes, these errors of judgment can impair leadership performance and distort safety-related decisions. Fortunately, although we are by nature disposed to make these errors of judgment, they are not inevitable.

Usually, with the benefit of a high-quality root cause analysis, we come to see the alternative course we might have taken or other decisions we could have made that would have prevented the incident. We then realize we made judgments that were incorrect—and preventably so. Such flawed judgments usually concern our assessment of future probabilities. If no adverse events have occurred within a particular service area for the past six months, does that mean the area is safe? If an adverse event happens for the first time, how likely is it a fluke?

After a series of adverse events, it seems clear what we should have decided and what course of action we should have taken. Nevertheless, if we look carefully at what we knew before the event, too often we had all the information we needed to make a safe decision, but we didn't pay attention to it. Why not?

The purpose of this chapter is to acquaint you with cognitive bias and help you devise strategies for avoiding the errors it can create.

This overview discusses findings relevant to healthcare and provides a framework for understanding cognitive bias from a practical, application-oriented perspective. A rich scientific literature[1] explores cognitive bias. The primary insight from the available research is that human beings tend to make predictably inaccurate judgments of known types about the probability of future events.

> Too often we had all the information we needed to make a safe decision, but we didn't pay attention to it. Why not?

Tragedy on Mount Everest

On May 10, 1996, five mountain climbers perished in an attempt to reach the summit of Mount Everest. Two were the world-renowned mountaineers Rob Hall and Scott Fischer, both skilled team leaders with extensive experience climbing at high altitudes. Hall was the leader of the Adventure Consultants expedition and had successfully guided almost 40 climbers to the summit over the previous six years. Fischer, leader of the Mountain Madness expedition, had reached the Everest summit only once, but he had a reputation as a skilled high-altitude climber. Climbing Everest has always been challenging and potentially dangerous: between 1922 and 1996, more than 160 people had died attempting to reach the summit.

---

1 *See, for example:*

Robyn Dawes, *Rational Choice in an Uncertain World* (New York: Harcourt Brace Jovanovich College Publishers, 1988).

John S. Hammond, Ralph L. Keeney, and Howard Raiffa, "The Hidden Traps in Decision Making," *Harvard Business Review*, 76 (1998): pp. 47–58.

Jonathan Evans, *Bias in Human Reasoning: Causes and Consequences*, Essays in Cognitive Psychology (London: Psychology Press, 1990).

Jerome Groopman, *How Doctors Think* (Boston: Houghton Mifflin Co., 2007). This highly readable and informative book deals specifically with cognitive errors in medical decision making.

Robin Hogarth, *Judgment and Choice* (New York: John Wiley & Sons, 1980).

Daniel Kahneman, Paul Slovic, and Amos Tversky, eds., *Judgment under Uncertainty: Heuristics and Biases* (Cambridge: Cambridge University Press, 1982).

Richard Nisbett and Lee Ross, *Human Inference: Strategies and Shortcomings of Social Judgment* (Englewood Cliffs, NJ: Prentice-Hall, 1980).

No single reason explains the 1996 tragedy; instead, a complex amalgam of causes, including deficient team functioning and flawed decision making, contributed to the outcome. The story provides an opportunity to study the way leaders shape behavior while balancing competing pressures within their teams or organizations. It shows how their words and actions influence perceptions and beliefs that can lead to hazardous behaviors. This case, described in *California Management Review,*[2] provides a particularly good explication of cognitive bias for our purposes because it occurred within a complex system encompassing multiple interactions, much like the ones facing healthcare safety leaders.

As background, the situation that prevailed prior to the Mount Everest tragedy can be characterized as follows:

- High-altitude mountain climbing was acknowledged as a challenging and potentially dangerous sport. Climbing Mount Everest was known to be particularly dangerous.
- The expedition leaders, Rob Hall and Scott Fischer, were highly skilled experts. They had exceptional reputations as guides and had a history of successful ascents at high altitudes.
- The clients (the participants in the climb) had invested significant resources (as much as $70,000 each) and expected to reach the summit of Everest.
- The participants had prepared for the physical rigors of the climb and were confident in their abilities.
- Agreed-upon guidelines shared by the Adventure Consultants and Mountain Madness expeditions helped to manage risk—e.g., the "two-o'clock rule" stated: "If you're not at the summit by 2:00 P.M. at the latest, you must turn around."
- There had been a recent history of favorable climbing weather.

Of the dozens of climbers ascending the summit that day, only six reached the top of Everest by 2:00 P.M. Around that time, four

---

2 Michael A. Roberto, "Lessons from Everest: The Interaction of Cognitive Bias, Psychological Safety, and System Complexity," *California Management Review*, 45 (Fall 2002): pp. 136–158.

of the remaining climbers abandoned their bid for the summit in order to begin their descent. Hall and Fischer, however, continued toward the summit with 15 others. Some of these climbers arrived as late as 4:00 P.M. When unexpected bad weather developed, poor visibility overcame the climbers, resulting in five fatalities between the two expeditions.

Cognitive bias played a significant role in the fatal decisions. Three types of bias were particularly noteworthy:

- *Overconfidence bias:* A tendency to overestimate the accuracy of your predictions, despite evidence to the contrary. Both the guides and their clients exhibited this bias: they felt confident about their success even though the data on successful ascents of Mount Everest did not justify it.
- *Sunk cost bias:* A tendency to make choices that support past decisions and escalate your commitment to a course of action in which you have invested, even when data indicate you may be on a losing course. After investing so much in the climb, the guides and clients did not follow the two o'clock rule to which they had previously agreed.
- *Recency bias:* A tendency to pay more attention to data that are recent and easy to recall. The recent history of very favorable climbing weather overshadowed the known probability of violent storms.

Cognitive bias and healthcare safety

Healthcare leaders can make most decisions without a purposeful consideration of cognitive bias. Cognitive bias can be disastrous, however, at critical decision points. Therefore, it is important to appreciate the role cognitive bias can play. Cognitive bias can cause healthcare leaders to underestimate exposure risk and overestimate the capability of systems to mitigate hazards. While any single decision may be insignificant by itself, a series of small, incorrect decisions may create a path to disaster.

Understanding cognitive bias improves decision making. Knowledge of cognitive bias positions the leader and treatment team members to intentionally question their own thinking and to isolate biases that increase hazards caused by poor decisions. Understanding the various kinds of cognitive bias helps leaders give due consideration to them when making decisions—especially crucial ones.

Human beings are not infallible, in part because of how we process information and make decisions. We are prone to cognitive biases because of how certain brain mechanisms work to help us accomplish two important tasks: 1) making the complex world simpler and more predictable; and 2) understanding new information by making it consistent with past information.

Cognitive biases are automatic and unconscious. You can gain a sense of their automatic, unconscious quality by imagining for one uncomfortable moment that you are a patient with a serious illness that often turns out to be fatal. Sitting across from you at her desk, your doctor is asking you to sign a consent form for a proposed treatment that is especially unpleasant and carries some risk of making your condition worse. Suppose the doctor says, "In cases like yours, the treatment is successful more than one third of the time." Research[3] shows that if she phrases the information that way, you are more likely to consent than if she says, "In cases like yours, the treatment fails nearly two thirds of the time." The facts are the same in the two statements. The difference lies in the automatic ways we all process the descriptions. Hearing "more than one third of the time" the treatment is "successful," we typically accept the risk and ask for the treatment. Given the identical facts stated as a "nearly two thirds chance of failure," we usually decline the treatment.

Although cognitive biases are usually not conscious, in the sense of immediately present to our conscious awareness, they are not *un*conscious in the Freudian sense: there is nothing blocking our awareness of them. Normally we attend to *what* we perceive and think. We pay less attention to *how* we perceive and think. When we do so, we can become directly aware of cognitive biases.

Cognitive biases reduce the sense of uncertainty, even though uncertainty is a central reality in medical decision making. Every

---

3 This illustration is adapted from Groopman, *How Doctors Think*, p. 242.

day, healthcare leaders make decisions under such conditions. Clinical and administrative leaders must satisfy multiple constituencies and meet budgets and schedules while simultaneously managing the risks inherent in the working interface. Medical leaders make decisions in the midst of life-and-death crises beset with uncertainty while facing the daily demand—from the law, their professions, their consciences, and their patients—to be completely competent and in full control. While cognitive biases may distort how we perceive the world, they do help us move forward in difficult situations.

Many cognitive biases have been identified, and some cognitive biases are known in the literature by different names. Thus, their definitions may be difficult to distinguish from one another. Below are biases that are important for patient safety and organizational culture change. They are organized (somewhat artificially) under two headings: biases revealed in the action of selecting data and biases shown in the use of data after its selection.

As you read through the descriptions, do not try to memorize every cognitive bias. Rather, study how these biases function in decision making. Look for how they enable a person to move forward in the face of uncertainty and how they can also produce errors of judgment.

Biases of data selection

Biases of data selection (summarized alphabetically in Table 7–1) enable pattern recognition by means of selecting some pieces of information and discarding others in order to establish a recognizable pattern. These biases differ from each other primarily in the (usually unconscious) principle by which we make these selections.

Prototyping

Prototyping is a tendency to automatically select data that fit a mental model of the situation. The successful practice of medicine has its roots in highly refined prototyping, which is a form of pattern recognition. A clinician has amassed years of experience with numerous illnesses and stored this experience as mental models of disease states. When she sees a patient, relevant models jump to

TABLE 7-1. BIASES OF DATA SELECTION.

BIAS	TENDENCY
Anchoring	Putting too much weight, or "anchoring," on the first information received or on a specific piece of information over all others; allowing the initial or preferred information to anchor subsequent judgments.
Confirmation bias (or selective perception)	Seeking and highlighting data that confirm your preconceptions, views, expectations, fears, or desires.
Fact/value confusion	Regarding and presenting strongly held values as facts instead of personal values.
Negative (or silent evidence) bias	Ignoring or discounting the importance of negative findings or missing data while giving credence to positive findings or the data at hand.
Prototyping	Automatically selecting data to fit a mental model of the situation.
Recency (or availability/ nonavailability) bias	Paying more attention to data that are easily available (e.g., most salient or recent, and therefore most memorable) while neglecting less readily available data.
Retreating from zebras bias	Ignoring the rare possibility when weighing the evidence. When you hear hoofbeats, you think about horses, not zebras.
Sample bias	Placing high value on a small sample that may be flawed due to inadequate sampling technique.
Search satisficing bias	Prematurely terminating the search for explanatory data because you have found something.
Status quo bias	Filtering information so as to favor the current situation.
Sunk cost bias	Making choices that support past decisions or escalating your commitment to a course of action to which you have invested time, energy, reputation, or money—even when data indicate the course of action may be a mistake.

mind. This expert thought process usually serves the physician and patient well, because it enables the doctor to make a rapid, efficient, and expert appraisal of the patient's malady. Prototyping can also lead, however, to a misdiagnosis or the selection of an incorrect treatment option when the patient represents an exception to the pattern embodied in the model. The most astute experts use models only as a starting point. The process of gathering data, evaluating it, and letting it evoke models of the illness is iterative. If the process is halted too soon, or if other biases interfere with data gathering and evaluation, serious errors can occur.

Prototyping problems are also common on the administrative side of the healthcare house. Consider the multidimensional design of many performance management systems: the performance criteria for individual and team functioning often include a dozen or more measures of service quality, timing, cost, and (occasionally) outcomes. The underlying belief—whether conscious or unconscious—is that "what gets measured gets done." When performance fails, leaders often look first to the measurement system: "We must be measuring the wrong things!" or "We must not be giving enough weight to the quality measure."

These diagnoses often emerge from leaders who themselves frequently: 1) grant undue forgiveness for performance failures; 2) change their performance expectations unilaterally and only later inform the workers responsible for fulfilling those expectations; and 3) formulate largely ungrounded assessments about the commitment of workers—clinical, technical, and administrative—to perform their work properly.

Among the problems to which these leaders are blind are the very unpredictability of their own behavior and the resultant erosion in trust between leaders and followers. These developments are detectable with the Organizational Culture Diagnostic Instrument (see chapter 3); they show up as adverse changes in the OCDI dimensions of procedural justice (team members see the leaders' behavior as unfair and untrustworthy), perceived organizational support (of team members by their leaders), and upward communication (by team members to their leaders about unsafe acts). The hazards and costs of cognitive bias thus often run deep in both the administrative and clinical chains of command in the healthcare organization.

Anchoring

Anchoring describes the tendency to give disproportionate weight to the first information received or to a specific piece of information over all others, allowing the initial or preferred information to "anchor" subsequent judgments. Where do you start when you are putting together the pieces of a jigsaw puzzle? You start with the first puzzle piece that catches your eye, whether because of its color or shape. Then you search for the other pieces that fit it.

When faced with an adverse event or when discussing what to do about a safety problem, how often does the first idea for which there is some evidence capture the organization's attention and come to dominate subsequent thought, discussion, and action to the exclusion of other causes, such as cultural and systems issues?

Expert diagnosticians generate a differential diagnosis (or said another way, a discrete risk assessment) within a few moments of seeing a patient. Although this is a productive and reasonable way to proceed, anchoring often biases the reading of subsequent information in favor of the first diagnosis that comes to mind and influences the physician to give insufficient attention and thought to other possibilities.

Recency (or availability/nonavailability) bias

Recency is the tendency to rely on those data that are most readily available (e.g., most salient or recent) and therefore easiest to remember, while neglecting less readily available data. Availability bias and anchoring are the most frequent biases at work in hectic, high-pressure environments such as the emergency room.[4]

---

4 Recency bias is frequently and clearly observed in the relationship between the senior leaders of healthcare organizations and their outside consultants. So well known is this phenomenon of remembering best the consultant who presented most recently (i.e., last in a series of competitive presentations) that consultants often go to extremes to be the last presenter. Among such ends that we have encountered in competition with other firms include: 1) reports of unexpected illness among the presenting team in order to request a later date; 2) in one case, a false report of a missed or cancelled flight to accomplish a postponement to a later date presumed to be the last presentation; and 3) claims of an automobile accident prohibiting a team from appearing at the appointed time. These likely false and unethical maneuvers have become more common than one might expect, because the recency bias of prospective clients is so strong and predictable. Although our evidence is at best anecdotal, we would estimate that the last consultant to present in a competitive bidding process wins the engagement half or more of the time.

Negative (or silent evidence) bias

Negative bias, which is related to recency, is the tendency to ignore or discount the importance of negative findings or missing data while giving credence to positive findings or data in hand. Another way to put together a puzzle is to select the puzzle pieces that are most readily at hand. For instance, the untrained root cause analysis will see most adverse events as related to personal error, since such errors are most salient.

If a physician has recently made a diagnosis, the probability of his making that same diagnosis in the next situation increases for no other reason than that he recently made the diagnosis. Similarly, if an administrator has been sued for the outcome of one of her decisions, she is likely to change her future relationship with the data that led to that decision, and she runs the risk of becoming overly aggressive in interpreting the data the next time around.

Retreating from zebras bias

The opposite side of the availability coin is neglecting data that are not readily available, which is called "retreating from zebras"[5]—the tendency to ignore the rare possibility when weighing the evidence. When you hear hoofbeats, you think about horses, not zebras. The problem, of course, is that sometimes the hoofbeats belong to zebras. If medications have been administered correctly every time for months, with no reported errors, the pharmacy manager feels little motivation to investigate the possibility that the system is flawed, and he may discount evidence that such flaws exist (or even punish the bearer of such evidence).

Confirmation bias (or selective perception)

Confirmation bias is the tendency to seek and highlight data that fit your preconceptions or confirm your views, expectations, fears, or desires. In patient safety, this bias most often shows itself in how we filter data to support our opinions about the need for systems change, how we conduct risk assessments, and how we approach root cause analysis. For example, selective percep-

---

5 Zebra retreat was named by Pat Croskerry in his excellent paper cataloging common cognitive errors in medicine, "Achieving Quality in Medical Decision Making: Cognitive Strategies and Detection of Bias," *Academy of Emergency Medicine* (2002) 9:11.

tion is implicit in the statement, "We don't have a patient safety issue here; we haven't had a sentinel event in more than a year." Such a statement is fallaciously based on the wrong measure—i.e., outcome data rather than upstream measures of exposure to hazard.

It is natural to believe that we are objective and don't "lie" in this way, but remember—these biases are universal and unconscious tendencies. They do not stem from intellectual dishonesty. This is why, using the scientific method, we must go to such great lengths to protect against cognitive bias.

Sunk cost bias

Sunk cost bias is the tendency to make choices that support past decisions and escalate our commitment to a course of action in which we have invested time, energy, reputation, or money—even when data indicate this course of action may be a mistake. This bias—clear in the Mount Everest case—manifests itself as a lack of flexibility; i.e., a difficulty recognizing when to change course. For example, a leader may not comprehend that the safety data are flashing a warning signal about the course of action to which she previously committed.

Sample bias

This bias—a tendency to place high value on a small sample that is flawed due to inadequate sampling technique—underlies much of the argument about evidence-based medicine and best practice guidelines. Those who strongly support the evidence-based/best practice approach see those who resist it as tending to rely too heavily on their individual clinical experience.[6] Another place where sample bias frequently shows up is in the exclusive use of outcome measures as a proxy with which to monitor month-to-month or quarter-to-quarter exposure. Outcome data usually provide an inadequate sample of exposure, because severe adverse events

6 There is nevertheless a legitimate flip side to this story. For a humorous example, see Gordon C. S. Smith and Jill P. Pell, "Parachute Use to Prevent Death and Major Trauma Related to Gravitational Challenge: Systematic Review of Randomised Controlled Trials," *British Medical Journal*, 327 (2003): 1459–1461. One might formulate a new cognitive bias that also represents the other side of this argument. The technology bias: the tendency to place unduly high value on data that lend themselves to algorithmic problem solving and thereby diminish the need for complex or emotionally difficult decision making.

occur relatively rarely and there are many intervening variables between exposures and adverse events.

Status quo bias

Status quo bias is the tendency to filter information so as to favor the current situation. Administrators, physicians, nurses, and other employees may be subject to this bias when considering whether to support organization and culture change. This preconception is another bias that often underlies the objection, "We don't have a patient safety problem here."

Fact/value confusion

Fact/value confusion describes the tendency to regard and present strongly held values as facts. For example, the debate over abortion, pro and con, often proceeds with value commitments playing the role of fact. In another example, a doctor may be unaware of how his personal religious or sectarian convictions influence how he interprets data and directs patient treatment.[7] A patient safety leader may make the opposite mistake, appealing only to facts to motivate a needed change in culture while remaining blind to or mute about the relevance of ethical and emotional issues.

Search satisficing bias

This cognitive bias is a tendency to prematurely terminate the search for explanatory data because you have found something. At the extreme, what you have found may be totally irrelevant, but it is reassuring to have a definite finding in hand, and its irrelevance may not be appreciated. More usually, the finding may be relevant but by itself insufficient. In any case, the finding is taken to be reason enough to terminate the search and take action. For example, do you stop looking when you find "human error" in a root cause analysis, without exploring what situational, systems, cultural, and cognitive biases set up those "at fault" to make the error? In unenlightened organizations, the issue may be dropped prematurely once blame is reasonably placed.

---

7 See, for example, survey research by University of Chicago physician/ethicist Farr Curlin. Farr A. Curlin et al., "The Association of Physicians' Religious Characteristics with Their Attitudes and Self-Reported Behaviors Regarding Religion and Spirituality in the Clinical Encounter," *Medical Care*, 44 (2006): pp. 446–453.

Biases of data use

These biases (Table 7–2) guide and distort the way we process or analyze data. They may compound the errors created in our selection of data.

TABLE 7-2. BIASES OF DATA USE.

BIAS	TENDENCY
Actor-observer bias	Explaining others' behaviors more in terms of their personalities than their situations, but doing the opposite when explaining your behaviors.
Commission bias	Taking action before you have a sufficient understanding of the situation.
Déformation professionnelle	Interpreting data and analyzing the situation from the perspective of your own profession while neglecting or discounting other perspectives.
Detached concern and dehumanization	Disconnecting emotionally from another person in order to help him or her—which has both adaptive and maladaptive applications. In its maladaptive form, this bias becomes dehumanization.
False consensus effect	Overestimating the degree to which others agree with you.
Group attribution error	Skewing your evaluation of data in favor of your group identification—whether defined by race, gender, belief system, or in other ways. A setup for in-group bias.
Hindsight bias	Filtering your memory of past events through current knowledge. Seeing past events as being more predictable than they in fact were. ("I knew it all along.")
Illusion of transparency	Overestimating your knowledge of others and others' knowledge of you.
In-group bias	Giving preferential treatment to members of your own group.

BIAS	TENDENCY
Momentum bias	Carrying forward with an existing course of action despite incomplete or even contrary evidence.
Negativity/positivity effect	When evaluating the causes of the behavior of a person you dislike, attributing positive behaviors to the situations surrounding the behaviors and negative behaviors to the person's inherent disposition. The inverse is the positivity effect, which comes into play when evaluating the causes of the behaviors of a person you like. Both are attributional errors.
Optimism bias	Being overly optimistic about the success of a course of action to which you have committed.
Overconfidence bias (epistemic arrogance)	Overestimating your abilities and the accuracy of your predictions, perceptions, and judgments, despite evidence to the contrary.
Projection bias	Assuming (unconsciously) that others resemble you in critical respects—thoughts, positions, values.
Rosy retrospection	Looking back and remembering the "good ol' days"; rating past events more positively than you rated them at the time.
Self-serving bias (Lake Wobegon effect)	Skewing your evaluation of data in favor of your ego (at the expense of others) as a means of protecting your self-image. Recalling your successes (e.g., with a particular procedure) more readily than you recall your failures. Believing yourself to be above average.
Wishful thinking bias	Overestimating the probability of good things happening. Preferring a particular course of events because its outcome is desired.

Commission bias

Commission bias is the tendency to act before we have a sufficient understanding of the situation. While we sometimes act precipitously because of the overconfidence bias (see below), we probably do so more often in response to pressure from the patient or from the situation to "do something." For example, we may inappropriately prescribe an antibiotic for a "sore throat," or make a new rule or procedure following an adverse event without first reaching a deep understanding of the systems that caused the event. Personality may also exacerbate this tendency: some of us are more predisposed to action than others. Those leaders or clinicians who are by nature less risk averse may take action more precipitously than those who are not so risk averse.

Déformation professionnelle

Déformation professionnelle is a French phrase—a pun on the expression "formation professionnelle," meaning "professional training." It describes the tendency to see things—and to interpret the data we have—from the perspective of one's own profession while neglecting or discounting other, broader perspectives. This bias can create conflict between constituencies. Building alignment requires that we successfully counter this bias. The intrinsic value of human life provides an adequate basis upon which to build a pan-organizational perspective for patient safety.

Momentum bias

Momentum bias is the tendency to carry forward with an existing course of action despite incomplete or even contrary evidence. In culture change undertakings, this bias can play out as a lack of responsiveness and flexibility. Does the leader try to fix the problem using more of the same (i.e., using an old approach that has momentum in the organization but has proved insufficient in the past)? Or does she discover a new approach?

Self-serving bias, actor-observer bias, and group attribution error

This set of biases skews our evaluation of data in favor of our ego (at the expense of others) as a means of protecting our self-image. These biases protect us, but they also undermine our ability to be responsible, and result in an overestimation of our abilities—a perfect setup for errors. At the same time, they lead to our responding to others' behaviors with blame rather than understanding.

Zimbardo[8] usefully points out that these biases have the effect of making each of us think that we are an exception: although biases snare *others,* they certainly do not snare *us*. Like the children of Lake Wobegon, we are all above average. Consider the implication seriously: It means that correcting one's own biases and avoiding the errors to which they predispose us poses a difficult challenge. It also means that if we are willing to listen, others can help us, because they do not share our bias in favor of ourselves.

As you read about the biases in this chapter, observe your own tendency to think how you would successfully outwit them. Try to set that bias aside and imagine yourself as truly vulnerable to them.

Attribution error biases operate through the tendency to understand one's own success in terms of personal powers and abilities while seeing one's failings as the result of bad luck or external, situational causes. Conversely, we also suffer from the tendency to explain others' failings in terms of dispositional factors, i.e., inter-

As you read about the cognitive biases, observe your own tendency to try and outwit them. Try to set that bias aside and imagine yourself as truly vulnerable.

nal factors such as personality or an inadequate sense of responsibility. For example, if someone else—particularly someone you do not like—gets in a car accident, it's probably because he is such a bad driver. If you get into a car accident, it's because the roads were slippery. This bias is self-serving because it excuses our own failings while blaming others for theirs. It leads immediately to (spoken or unspoken) blame as the first response to a problem.

At the group level, the group attribution error acts in the same way to protect my group identifications from members of other

---

8 Philip Zimbardo, *The Lucifer Effect: Understanding How Good People Turn Evil* (New York: Random House, 2007), p. 447.

groups—whether defined by race, gender, belief system, or in other ways. This phenomenon creates the in-group bias (below).

Hindsight bias

Hindsight bias is the tendency to think that past events were more predictable than in fact they were. This bias leads to analyzing errors in terms of blame and, with the benefit of hindsight, counting as preventable those events that were not. We tend to feel that had *we* been the person involved, we would have foreseen the bad outcome and made a better decision.

Rosy retrospection

This is the tendency to look back and remember the "good ol' days." We recall our successes with a procedure or system more readily than we recall our failures. This predisposition distorts our judgment when we contemplate giving up a familiar procedure in favor of a new one such as an evidence-based best practice. It may even blind us to the failure of the familiar procedure to produce the outcomes we ascribe to it. It is part of what motivates resistance to change.

Overconfidence bias (or epistemic arrogance) and optimism bias

Clinicians and nonclinicians alike tend to overestimate the accuracy of our predictions, perceptions, and judgments as well as the strength of our abilities, even in the presence of evidence to the contrary. Similarly, we tend to be overly optimistic about the success of a course of action to which we have committed. In patient safety, these biases may show up as an inability to see the warning signs that an intervention is not working. Perhaps, for example, the administrator assumes that because he has deployed the intervention, the organization is following it, and he fails to notice the telltale signs that it is not. These biases can also color how we estimate an intervention's likelihood of success.

In-group bias

This is the tendency to give preferential treat to members of one's own group. This bias strengthens subculture alignment at the expense of alignment with the larger culture. For example, it may pit doctors against nurses or nurses and doctors against the administration. In an unhealthy culture, in-group bias can defeat the successful investigation of adverse events when those involved act to protect group members.

Illusion of transparency, projection bias, and false consensus bias

These biases describe the tendency to overestimate your knowledge of others and others' knowledge of yourself, to believe that others resemble you in critical respects (e.g., the values they hold), and to overestimate the degree to which others agree with you. These biases create the illusion of understanding and thereby undermine motivation to identify and resolve differences. They set up premature judgments and misunderstandings. In clinical situations, they lead to communication errors, such as the assumption that the patient understands your instructions just as well as you do. In the leadership of culture change, these biases precipitate premature action: acting before there is true buy-in for the change.

Wishful thinking

Wishful thinking means to prefer a particular course of action because the outcome is desired. This bias is just one of many ways that emotions can distort judgment.[9]

Negativity effect and positivity effect

We know that dislike for a patient can easily incline treatment team members to errors of judgment. Disliking a patient can predispose team members to avoid the patient, not listen to him, cut his communication short, and be insensitive to the significance of his complaints. Groopman points out that we don't like patients

---

9 While it is true that emotion can distort judgment, emotion is also necessary for competent judgment. See, for example, Antonio R. Damasio, *Descartes' Error: Emotion, Reason, and the Human Brain* (New York: G. P. Putnam's Sons, 1994).

who appear to us to be uncooperative with the treatments we prescribe, e.g., patients with hypertension or diabetes who don't take their medications, patients who are morbidly obese who won't exercise and diet, and patients addicted to cigarette smoking or alcohol who "don't care about themselves" enough to stop consuming the substances. We also tend to dislike patients whom we fail to treat successfully, those who are demanding, and those who complain incessantly or have psychological problems. All these patients are more vulnerable to becoming victims of medical errors.

But positive emotions can blind us as well. This is the truth behind the adage that a doctor who treats himself has both a fool for a doctor and a fool for a patient. Groopman[10] cites one of his own cases, involving a patient he liked very much who was being treated with Adriamycin for osteosarcoma. The patient became exhausted and developed a fever of unknown origin. During the workup, Dr. Groopman avoided adding to the patient's discomfort by choosing not to examine the patient's buttocks as a possible source of infection. Unfortunately, later that day the patient developed bacteremia and septic shock from an abscess on his buttock. A common leadership example is selecting a change agent that you like rather than one who has demonstrated the skills, knowledge, and ability required for success.

The negativity or positivity effect comes into play when we attribute different motives to others' behaviors depending on whether we like or dislike them. We might assume, for instance, that a likable patient didn't take her medications properly because she was not given adequate patient education (the situation, not her personality, was at fault). But when a patient who is unlikable forgets to take his meds, we tend to interpret this lapse as a personal failing—a character defect.

Detached concern and dehumanization

Detached concern means to disconnect emotionally from another person in order to help him or her. This interesting, emotionally motivated bias shows clearly how a single bias can function both adaptively and maladaptively. Healthcare professionals and organizational leaders use this bias adaptively—for instance, when a

10 Groopman, *How Doctors Think*, pp. 48–54.

leader dispassionately studies performance statistics or a surgeon temporarily ceases to regard the patient as a person in order to cut the patient open.

When detached concern is not used for the patient's benefit, it is called dehumanization—the psychological act of robbing others of their humanity. Dehumanization appears in healthcare when a leader discounts the emotional significance of the unnecessary human suffering implicit in the organization's safety performance statistics or when a frustrated doctor describes a patient for whom she can find no adequate diagnosis or treatment as "a crock."

Habitually using dehumanization to protect oneself emotionally can set up a leader for the charges of not listening and arrogance, which lead to both increased likelihood of errors and increased malpractice exposure—for both the hospital and the physician. Healthcare professionals and administrators are particularly prone to dehumanizing patients when working in the midst of financial crisis or struggling to keep up with an overwhelming caseload, neither of which is an infrequent occurrence in healthcare today.

Case study: Cognitive bias in manufacturing

To understand how cognitive bias can work below the level of your awareness and affect your leadership behavior and your influence on others and the culture, consider the following case from a manufacturing plant. Focus on the plant manager's cognitive bias and resulting behavior. Next, look for the effects of his behavior on his direct reports. Finally, look for the biases active in his team. Notice that none of these people were aware of the role the biases were playing in their behavior:

> The manufacturing facility is planning a major expansion in six months during the slow production season for its cyclical product, at which time it will be replacing an aging production line with completely new equipment. The annual safety audit of the plant recommends the installation of emergency shutdown interlocks—safety devices that automatically shut down the system when triggered by an out-of-bounds condition or an

inappropriate employee action. This change has prevented serious injuries at similar plants in the industry. However, because of the age and complexity of the existing equipment and the design and layout of the plant, installing the new interlocks would require shutting down the line for at least a week, and the line is running around the clock to meet market demand during this busy season. Gross sales amount to around $1.3 million per day.

The plant manager reviews the results of the safety audit with his manufacturing, maintenance, and safety managers. He explains the problems that would arise if he shuts down the line and misses production targets. The safety manager agrees with the audit recommendation but reports that this plant has never had the type of incident that the interlock would prevent. The manufacturing manager and maintenance manager agree that meeting the production target will be impossible if the line is shut down, and since the line will be shut down in six months anyway, they both recommend that the installation of the safety device be deferred.

Three weeks later, a fatality occurs. The death of the employee would have been prevented had the leaders shut down the line and made the recommended changes.

Here are some of the biases that contributed to this fatality. The plant manager created bias among his reports by acting out his own anchoring bias. He *anchored* their consideration of the problem by opening the discussion with a review of the problems that a decision to shut down the line would cause. This approach created a frame of reference that effectively served as a roadblock to exploring the consequences of not shutting down, and it created momentum in the minds of his reports toward postponing the shutdown. Moreover, the known, certain costs of shutdown were allowed to hold more sway than the uncertain cost of not shutting down—the retreating from zebras and silent evidence biases. All the managers succumbed to the recency bias—they had operated this line safely in the recent past. In addition, they were all also probably influenced by both the status quo and wishful thinking biases.

Exercise: Identify the biases

In the following medical case history, identify (at the risk of activating your own hindsight bias) the cognitive biases that may be implicated in the decision errors:[11]

> The patient, a 30-year-old female who had converted to a religious faith forbidding blood transfusions, sustained life-threatening injuries in an automobile accident. The emergency room clinical staff judged that a blood transfusion was a necessary life-saving measure, but the patient declined to be transfused because her religion forbid it. The staff judged her competent and was prepared to honor her refusal.
>
> When the family and clergy arrived in the emergency room, the family disagreed with the decision not to transfuse. The family argued that the patient had converted to the religion only a few weeks prior to the accident and did not really have time to sufficiently understand the implications of her act. Her minister denied this, saying that she fully understood the implications of her decision to convert, including, specifically, the ban against transfusions.
>
> Shortly thereafter, the patient lost consciousness. At this point the emergency room staff reversed their decision not to treat, and they gave her a blood transfusion. Ultimately, the patient recovered from her injuries and was discharged.
>
> The patient and her minister then sued the hospital and the emergency room staff—and won. The court ruled that the patient's civil rights had been violated.

Why did the clinical staff decide to take action, after being explicitly told by a mentally competent patient that she declined certain kinds of care? Although we have no inside information on this case, we suspect that certain biases were involved. When the patient lapsed into unconsciousness, the overconfidence and confirmation biases may have caused the staff to judge that they were correct about the life-saving necessity of a blood transfusion. The *déformation professionnelle* bias may have limited the treatment team's frame of reference to that of healer and left them unable

11 Agency for Healthcare Research and Quality, "No Blood, Please," *Mortality and Morbidity Rounds on the Web*, May 2004. Viewed at www.webmm.ahrq.gov.

to consider the civil liberties framework in which they practice medicine.

The projection bias may have made it impossible for the treatment team to believe that the patient wanted what she had requested because it was not a choice they could imagine wanting themselves under the same circumstances. Wishful thinking may also have distorted the treatment team's judgment: they wanted to make the decision that would allow the patient to live.

Other emotional factors that may have biased the treatment team's decision were the family's pressure to save the patient and the team members' fear of a lawsuit if they didn't take steps to save the patient. Finally, the team may have disliked the clergyman or his beliefs about transfusion.

Again, notice that these biases operate primarily at an unconscious level. Had the treatment team members attended to how these biases were shaping their actions, they would have had more choice in the matter.

Putting your knowledge to work

How does the successful safety leader put this knowledge to work? During important decisions, successful decision makers use their working knowledge of cognitive bias to self-monitor and to enlist others in the effort to monitor for cognitive bias. For the treatment team member and safety leader whose choices affect the success with which exposure to hazards is controlled, this exercise can be critically important.

Frame the issue

- First, consider the problem on your own, then gather input from others.
- Don't be stampeded. Resist unreasonable pressures to make immediate decisions. Cultivate the ability to approach emergencies and other high-pressure situations with a degree of mental calmness.
- Consider how the problem has been framed by yourself or others. Does it invoke *déformation professionnelle* or other

strong, limiting frames? What history does it bring? What assumptions? How much momentum?

- Seek to widen your frame of reference and look at the problem with fresh eyes.
- Ask others how they are inclined to frame the issue.

Select the data

- Consider the prototypes the situation invokes, and look for other mental models that might be important to consider, especially models whose neglect represents a serious error.
- Discuss with others the data they would consider; seek data from a variety of sources; and intentionally attend to data that discount your theory.
- Consider whether you are anchored in the wrong data.
- Ask yourself what forms the basis of your assessment. Do you have valid and applicable data, or are you acting on recency, selective perception, or status quo bias? Are you confusing facts and values?

Judge available options

- Ask about alternative choices, interpretations, and decisions, and define competing options clearly. The status quo is never your only option, and your choices are rarely binary (i.e., to act or not act). In important decisions, always generate a short list of alternatives.
- For all decisions with history, verify that you are not giving undue consideration to sunk costs.
- Identify a credible team member or other person to play your devil's advocate.
- Ask yourself what your emotions are telling you to do and whether their guidance is appropriate.

Learn from your mistakes

- Study the research on cognitive bias and encourage other safety leaders to do the same.
- Seek help from colleagues to dissect erroneous decisions to uncover the cognitive errors involved, their triggers, and what might have prevented the errors.

Most important, the successful leader understands that the culture influences the extent to which errors of cognitive bias are allowed to flourish. A culture in which open communication is encouraged and mistakes are openly shared and dissected is the best protection against the potentially detrimental effects of cognitive bias.

Does the organization sustain a value for open and free-flowing communication? (Review the upward communication and approaching others dimensions of the OCDI in chapter 3.) In such an environment, people feel free to say what is on their minds, to speak up and challenge the assumptions that underlie biases. Does the culture actively encourage such dialogue, and do leaders both support and model it? Zimbardo cautions that organizational systems "create hierarchies of dominance with influence and communication going down—rarely up—the line."[12] Upward communication, especially after a safety breakdown, needs continual encouragement.

Are there traditions and systems for the detection and blame-free exploration of biases and errors, and for sharing lessons learned? These cultural attributes and practices are fostered by a strong transformational leadership style (see chapters 4 and 5) and reflected in high scores on the OCDI's four organizational dimensions (procedural justice, perceived organizational support, leader-member exchange, leadership credibility).

Which of the seven leadership best practices provide the greatest protection against falling victim to cognitive bias, and which foster a culture that guards against bias errors? Do all your leaders have high Leadership Diagnostic Instrument (LDI) scores for these practices?

---

12 Zimbardo, *Lucifer Effect*, pp. 10–11; see also pp. 445–446.

Let's go back to the case of the manufacturing plant that declined to install emergency shutdown interlocks, with fatal results. How could the decision making have been different? Had the plant manager been aware of the cognitive biases at work, he could have put the question on the table without anchoring. He could have asked for data on the extent of exposure (as opposed to incidents) at his plant versus other plants that had reported incidents, and on alternative control methods. As the group narrowed in on a decision, he could have asked someone to argue for an alternative. Perhaps most important, long before this discussion ever arose, he could have fostered a culture in which each of his subordinates felt encouraged to point out possible cognitive traps during the discussion of any important decision.

Exercise: Freedom from bias

The first step in protecting your decision making from cognitive bias is to become aware of the biases themselves. The second step is to become aware of when you are likely to be subjected to them. Divide a piece of paper into two columns. Review the list of cognitive biases in Tables 7–1 and 7–2. For each one, write down in the left-hand column the most likely circumstance under which you would fall subject to that bias. Using the above recommendations for coping with bias, write down in the right-hand column ideas on how you might avoid succumbing to cognitive bias errors in each of the circumstances you identified.

Encourage bias detection

Each bias by itself can distort decisions. Even more dangerously, more than one bias may be involved in a single decision and serve to intensify the distortion—e.g., a hastily made decision in response to emergency room overflow that then becomes accepted and difficult to change because of the status quo bias. Or perhaps a head nurse makes a poor decision about nurse overtime in response to the emotional demands of a specific situation, but then continues to implement and defend the decision because of the sunk cost bias—even though the scheduling decision has become manifestly inappropriate.

Bias can help explain why capable leaders make and accept poor decisions—why a skilled clinician will filter out hazards (overconfidence bias), why a senior leader will not terminate a poor performer he has coached, despite the team member's consistently poor performance (sunk cost bias), and why the treatment team will live with a safety system that delivers weak safety performance (status quo bias).

Despite our best intentions, the way we process information sets us up for errors of judgment. We now know that we can take steps to protect ourselves from such errors. Although each of us must work on this challenge ourselves, we need not work alone; a culture that encourages bias detection and examination provides perhaps our best defense against the problems these biases can cause. But how do we build such a culture?

In prior chapters we explored the five elements of the Blueprint for Healthcare Safety Excellence (leadership, organizational culture, working interface, organizational sustaining systems, and healthcare safety–enabling elements) and learned the importance of leadership to culture and of culture to safety performance. We examined the utility of applied behavior analysis and ABC analysis. We are now in a position to develop and implement a strategy for building a strong safety climate and leading culture change.

DESIGNING YOUR SAFETY IMPROVEMENT INTERVENTION

CHAPTER EIGHT

DESIGNING YOUR SAFETY IMPROVEMENT INTERVENTION

Recent research has shown that in some important respects, members of different geographic cultural groups think differently. For example, Asians and Westerners characteristically employ distinct cognitive processes. These processes and their differences have been well documented and thoroughly studied.[1] When an Asian moves to the West or a Westerner to Asia, the newcomer's cognitive processes begin to take on the characteristics prevalent in the new country of residence. How does this transformation happen?

Language acquisition poses a similar mystery: How do babies effortlessly learn their native tongue without being taught or reinforced, and seemingly without adequate exemplars?[2] A similar question pertains to the acquisition of moral standards. Children automatically adopt the moral sensibilities of the cultures in which they grow up.[3] How?

Imagine that you have moved to a remote country where you understand the language but where the culture is foreign. Gradually, you begin to fit in and understand the culture, usually without your being aware of the adjustments this change entails. By what process does your acculturation occur?

---

1 Richard E. Nisbett, *The Geography of Thought: How Asians and Westerners Think Differently...and Why* (New York: The Free Press, 2003).

2 Steven Pinker, *The Language Instinct: How the Mind Creates Language* (New York: HarperCollins, 2000).

3 Marc D. Hauser, *Moral Minds: How Nature Designed Our Universal Sense of Right and Wrong* (New York: HarperCollins, 2006).

Just as a person must transfer from one country and culture to another, all healthcare workers must transfer from a safety-optional to a safety-dominant culture. Safety is not a program—it's a way of life. Getting safety right requires not just everyone's willing participation but their active and wholehearted engagement and cooperation. In one sense, sustained engagement *is* culture change. Systematically reducing exposures to hazard in the working interface takes the active collaboration of leaders and team members. All employees have a central role to play, and they and their patients stand to benefit directly. Whether team members' engagement comes easily or with difficulty depends on the atmosphere created by leaders.

A leader creates organizational culture with his every thought, word, and deed. He sets organizational direction *explicitly*. He establishes priorities for the people who work for him and determines what they regard as important. Treatment team members' and employees' perceptions about what is important to the leader constitute the safety climate. As the leader maintains his focus and sustains a strong safety climate, the organization's culture gradually absorbs the climate's values and comes to support safety.

A leader creates culture with his every thought, word, and deed.

Furthermore, and perhaps more important, a leader sets organizational direction *implicitly*—without ever saying a word—by the signals her colleagues receive about her emotional and physical reactions to situations and from her decisions, behaviors, and leadership style. These nonverbal responses define the tone at the top as much as any overt, intentional action, and this tone quickly permeates the organization and establishes the feel of its culture. Haberdashers speak of the "hand" of a fine silk tie—i.e., the fabric's distinctive texture and firmness. Analogously, the culture of the organization stems from the feel of the fabric of relationships modeled by leadership behavior.

Employees observe and "set switches" based on a leader's explicit and implicit behavior; the leader creates culture because he provides behavior models. A leader sets organizational direction, which, in turn, prescribes employee behavior and defines what that behavior means; she creates culture through her vision. A leader shares his vision, telling people the meaning of their activity; he creates culture in his communications. A leader creates expectations and the standards by which behavior is judged; she creates culture through norm-setting. A leader provides employees with opportunities to become committed; he creates culture through engagement. A leader recognizes performance; she creates culture through acknowledgement and reinforcement.

For better or worse, a leader creates culture constantly in everything he says and does, both explicitly in his direction and implicitly through his decisions, reactions, behaviors, style, and body language—all of which shape his relationships with others in the organization, irrespective of the arrangement of boxes on the organization chart. A leader creates culture whether or not she intends to or is aware she is doing it. Creating culture is neither easy nor difficult for a leader to do; it is unavoidable.

The Leading with Safety process

Instilling specific new attributes into an existing culture requires a conscious, concerted, and sustained leadership effort. Leading with Safety is a highly effective,[4] two-phase, eight-step process that helps healthcare organizations build a healthy safety climate and a strong organizational culture that continuously improves safety performance (Table 8–1). Steps 1 through 4 (phase I) are discussed here; steps 5 though 8 (phase II) will be covered in the next chapter.

Precisely how these steps are undertaken depends on a variety of factors peculiar to the institution and its leadership: Does the healthcare organization have one location or many? How large is it? What is its organizational structure? Is the objective of the safety initiative to achieve broad improvement across the organization or to pinpoint certain areas?

Phase I of the Leading with Safety process belongs exclusively to leadership; much (but not all) of phase II can be delegated.

---

4 Thomas R. Krause, *Leading with Safety* (New York: Wiley & Sons, 2005) and Thomas R. Krause and Thomas Weekley, "Safety Leadership: A Four-Factor Model for Establishing High-Functioning Organizations," *Professional Safety* (November 2005): pp. 34–40.

Phase I starts by giving organizational and professional leaders (who should include the board, administrative leadership, and key leaders from nursing, the physician group, pharmacy, operations/ building management, and safety) a way to speak, understand, and consider how safety works in the organization.

TABLE 8-1. THE LEADING WITH SAFETY PROCESS.

Phase I: The Patient Safety Academy	
Step 1:	Gain leadership alignment on patient safety as a strategic priority.
Step 2:	Develop a patient safety vision.
Step 3:	Perform a current state analysis using the Blueprint for Healthcare Safety Excellence.
Step 4:	Develop a high-level intervention plan for Phase II, including specific outcome measures and related objectives.
Phase II: Achieving Safety throughout the Organization	
Step 5:	Engage the organization in the Leading with Safety process.
Step 6:	Realign systems, both enabling and sustaining.
Step 7:	Establish a system for behavior observation, feedback, and problem solving.
Step 8:	Sustain the Leading with Safety process to continually improve the organizational culture and patient safety.

After leaders gain an understanding of organizational safety, they are asked to commit to and align on the necessity of endorsing safety as both a strategic priority and an enduring value (step 1). In step 2, the leaders develop a coherent and tangible picture of what their organization, systems, processes, and people will look like as the organization achieves the level of safety to which the leaders aspire. Step 3 aims to achieve a comprehensive understanding of the prevailing current state of safety as indicated by current readings of key variables that define and determine the organization's level of safety performance. Finally, the leaders develop an intervention plan to strengthen these variables as needed (step 4).

Phase II (next chapter) adds additional detail to the leaders' plan to drive safety at the behavior level throughout the entire organization and down to the level of the working interface. Phase II establishes clear responsibilities for monitoring progress and for using new, continually generated information as a basis for continual problem solving. Although leaders can delegate much of the phase II work, success requires their ongoing involvement in many capacities: providing direction, focus, oversight, resources, and recognition as well as modeling new, safety-critical behaviors and best practices. The leaders' continued participation ensures that the organization stays the course.

Phase I: The Patient Safety Academy

The Phase I work can be accomplished rapidly (within 90 to 120 days) if the leaders are highly motivated and alignment is strong. Organizations differ in their needs; leaders differ in how fully they are willing to commit to patient safety and in how quickly they want to move. Depending on the organization's current state and the leaders' objectives, the duration of phase II may range from one to three years.

Although we describe the Leading with Safety process in steps, please be aware that altering culture for organizational safety is neither rigid nor routine. Culture change requires that leaders act with determination (sometimes uncompromisingly and unilaterally), but it also requires a nimble, flexible response. Different organizations proceed at different speeds, and some may need to spend more time on certain steps than on others.

Culture change requires that leaders act with determination—sometimes uncompromisingly and unilaterally.

Some organizations complete much or all of phase I using a retreat format, which enables leaders to disconnect sufficiently from day-to-day responsibilities and interruptions (including, especially, cell phones and e-mail!) to think strategically. High-quality thought can be encouraged with a degree of leisure, achieved by temporarily taking a break from the demands of the daily work environment.

Who should attend the first session of the Patient Safety Academy? It depends on the level of the organization initiating and sponsoring the change process. If the governing board originates or sponsors the process, then sponsoring board members (or the relevant board committee) and the CEO should attend. If the CEO originates the process, the CEO and the senior leadership team should attend. Ideally, in that case, the CEO invites board members as well, because it's important that the board and the executive and clinical leaders are aligned around the strategic importance of safety. In either case, the participation of both administrative and clinical leaders is essential to a successful start. The absence of a critical constituency from the patient safety retreat can send a disabling signal about the intended depth and breadth of the initiative to improve safety.

Step 1: Gain leadership alignment on patient safety as a strategic priority

Surprisingly, many organizations continue to approach safety as a compliance or licensing issue. This approach misses the underlying connection between safety and excellence, not to mention the underlying ethic of doing the right thing. Safety belongs among any healthcare organization's strategic objectives and on its short list of core values.

A great deal has been written about strategy; the word has become so hackneyed that it has lost much of its distinctive meaning. Strategy, to us, means the plan by which an organization intends to accomplish its final objectives, the objectives for which it exists.[5] Since it aims to actualize the organization's fundamental reason for being, the strategy needs to cohere around the organization's core values and guiding principles.

---

5 One text on this subject that has stood the test of time is *Top Management Strategy: What It Is and How to Make It Work* by Benjamin B. Tregoe and John W. Zimmerman (New York: Simon and Schuster, 1980).

The broad success of a healthcare organization depends on the attainment of patient and employee safety. In this sense safety is foundational to everything else the organization needs to do. If an airline exists in part to build enduring relationships with its customers and employees and to assure both constituencies about the reliability of its promises, its strategy must include a guiding principle such as Southwest Airlines' well-known mantra, "hire for attitude, train for skills." Creating a feeling of welcome among its customers and generating a feeling of mutual support among all its workers are two of Southwest's strategic priorities.

An organization's strategic priorities are of such central importance to its purpose and its very existence that the organization's leaders need at all times to have a clear and accurate picture of where the organization stands with respect to them. In the example just cited, Southwest Airlines needs (and seeks) regular feedback from employees and customers regarding the quality of their relationships with the company and the reliability of its promises.

Healthcare service organizations exist first and foremost for the purpose of minimizing suffering and promoting healing and health. Neglecting employee or patient safety *necessarily* increases ill health and suffering; protecting patient and employee safety provides a *necessary* means of achieving the organization's foremost purpose. For this reason, not to mention its ethical imperative, safety should be both a strategic priority and a core value for any healthcare service provider. And the organization's high-level, long-term plan for ensuring patient safety should be part of its overall strategy.

But healthcare is also a business, and it exists—even if it's a nonprofit corporation—to create sufficient profit (or surplus) to sustain the enterprise. Therefore, profitability is one of its strategic priorities. With more than one strategic priority, the possibility of conflict arises. Which objective commands primacy?

This potential conflict can cause lack of alignment among leaders and induce a suboptimal implementation of strategy. Moreover, conflicting departmental and professional commitments often fragment implementation efforts. Leaders need to face these hard choices[6] together and forge continuing leadership alignment on strategy and values. By facing them together, leaders are doing ethical work: using dialogue to weigh and choose among competing and conflicting goods. Neglecting this work carries the cost of any strategic or ethical lapse: suboptimal performance.

A healthcare organization's first responsibility

Ensuring freedom from accidental injury for patients and a safe workplace for employees should be a core value. It also should be the first responsibility of the healthcare organization's governing body and of its operational leaders. Accordingly, safety is both an ethical issue and a risk management issue for the board and the senior leadership to monitor continually, to oversee in a coordinated way, and to correct aggressively as needed. In the commercial arena, recent and ongoing research by the Corporate Executive Board (CEB), for example, identifies best practices related to risk assessment and ethics awareness. In these two areas CEB recommends quarterly reporting to the board of key risks and mitigation plans, plus the annual reporting of ethics survey results indicating degrees of compliance with the established code of conduct, benchmarking with comparable organizations, and careful analysis of survey differences between and among levels of leadership in the organization.[7] These suggestions have wide applicability in healthcare delivery organizations committed to ensuring employee and patient safety.

Are the leaders in your organization aligned on safety's strategic role and their responsibility for safety? What are their reservations or concerns? How can their concerns be addressed?

Setting the right tone at the top

"Tone at the top," the byword of compliance with the Sarbanes-Oxley Act, is an essential element in the creation of organizations with incident-free operations. (By *incidents* we mean increases in

6 See Tregoe and Zimmerman, *Top Management*, p. 119: "Top management's interest and involvement in the formulation of strategy must always be directed against the end results: a clear and useful statement of strategy which guides those choices that determine the nature and direction of an organization..."

7 Corporate Executive Board, *A Compendium of Board Reports: Enhancing Oversight of the Compliance and Ethics Program* (Washington, DC, 2006), p. 167.

exposures to hazard, only some of which result in patient harm.) Attention to safety in all its dimensions—including exposures or risk, and not just sentinel events—starts at the top. The top includes representatives of the organization's stakeholders, including patients, and not just senior leadership.

Setting a tone in the boardroom favoring performance means more than reviewing the safety performance statistics at each meeting and more than visiting patient care units. It means paying attention to the full picture of patient safety, requiring accountability, and expecting improved performance—without perpetuating a culture of blame. An attitude of diligence makes possible the improvement of leading safety indicators, the delivery of incremental hazard reduction, and the consequent enhancement of the organization's soundness and sustainability.

Are the leaders in your organization aligned on safety's strategic role and their responsibility for safety?

The safety tone is set at the top primarily by the care and astuteness of board-level and senior-level listening to the safety outcomes of the organization and to the upward communication from operating management about the safety climate. While organizational culture may take longer to change, effective listening and caring about patient safety almost immediately alters the safety climate and sets a tone for hazard avoidance.

The tone at the top is delicate. It can change subtly, unnoticed by those at the top. Ensuring safety requires continual reexamination of organizational tone as well as a continuous predisposition to listen to staff perceptions of changes in tone.

Do your leaders understand how they create the tone at the top? Do they understand how the tone relates to their strategic objectives? Are leaders committed to the requisite listening, and does the organization encourage the upward communication about safety that leaders need to listen to?

Gathering the right information

How does a healthcare organization know that its safety performance metrics are providing a complete picture of the right variables? A moderate-sized Midwestern hospital had virtually no significant events in three years, and its CEO was proud of its performance. Then "out of the blue" the hospital experienced a run of neonatal deaths. "Why didn't we see this coming?" the CEO demanded.

Effective governance and stewardship require timely and accurate data. As we write, neither the boards nor the senior leaders of most healthcare organizations in the United States have access to valid and complete information on the safety performance of their organizations. It is not that these data are kept from them; rather, most organizations have not invested in the development of the active monitoring and reporting systems needed. They rely too heavily on lagging data sources such as incident reporting systems and malpractice claims—both useful, but widely recognized as inadequate for assessing performance. Even the quality data that are becoming subject to public reporting provide an important but incomplete picture of patient and employee safety.

When the board and senior leaders of a healthcare organization become serious about the pursuit of safety for their patients and employees, they insist on regular updates on both upstream and downstream safety measures. These updates include:

- The percent conformance to proactive safety-critical behaviors and processes[8]
- The progress made on identified systems issues and hazards
- The rate of adverse outcomes, such as hospital-acquired infection rates, adverse drug events, patient complications, and mortality

These updates are organized to indicate whether overall performance is improving in line with organizational objectives.

---

8 Which implies previously having identified the inventory of safety-critical behaviors and having established regular means of observing and reporting their frequency of occurrence.

Are the leaders of your healthcare organization aligned on the necessity of such data? Which data are currently available, and which are not? What will it take to complete the safety data set?

How safety relates to other strategic objectives

Hospitals should look beyond the direct outcomes of accident prevention and ask: What is the relationship between safety and other performance metrics in this organization? Resources not consumed in worker compensation claims or treating avoidable patient complications can be redeployed to meet core organizational goals. Improved patient safety performance increases the efficiency of intensive care unit (ICU) bed use and increases throughput. Safe and reliable care lowers costs to payers and potentially increases market share. Effective patient safety practices increase staff loyalty, decrease staff turnover, and reduce associated hiring and training costs.

While the relationships between safety and other performance metrics are interesting from a theoretical point of view, we pose the question as an empirical matter. Understanding these relationships as they exist in your organization, at this time, provides data relevant to reconciling conflicts between your organization's strategic objectives.

Leaders should seek to determine the longstanding statistical relationships between variations in patient safety outcomes (e.g., risk-adjusted falls, adverse drug event rates, hospital-acquired infections, complications, preventable deaths) from month to month and quarter to quarter, and contemporaneous changes in staff turnover, length of patient stay, patient satisfaction, financial results, and the strength of staff loyalty to the organization. Our client experience suggests these statistical relationships are idiosyncratic to the operations of each facility—i.e., no two delivery organizations have identical patterns. Moreover, these unique relationships, when traced to root causes within a given facility, often reveal the organizational impediments to both safety and sustainable growth.

By examining the relationships between patient and employee safety on the one hand and organizational outcomes on the other, as well as the underlying causes of shortfalls in both, a hospital and its board can assess the contribution that safety makes to the organization's overall value—or the degree to which breakdowns in safety are inhibiting the sustainability of the delivery enterprise.

Do your leaders have the data they need to understand, monitor, and reconcile the relationships between safety and other strategic objectives? Have they anticipated the need for operating principles to guide choices among competing strategies?

The necessity of prevention

How does a healthcare organization know it is being preventive, and how does it measure hazards in the absence of adverse events? Virtually every patient injury is preceded by lower-level decisions and outcomes that increase the likelihood of a safety failure. The catastrophic outcome—a sentinel event, serious injury, or death—can be seen as the tip of an iceberg embedded in a larger architecture of behaviors, practices, and outcomes that made the greater loss predictable. Leading indicators of lower-level safety decisions reveal the organizational culture that gives rise to costly failures. Board members and senior leaders should ask which leading indicators are predictive for their organization, including measures related to organizational culture and safety climate. Then they should ask what is being done to move those leading indicators and how they are changing over time.

Board members and senior leaders should ensure that everyone in the organization fully understands the safety implications of what goes on in the working interface, where staff interact with patients and with the technology and infrastructure that support patient care. Ultimately, safety excellence depends on keeping the working interface free of hazards, which includes systems, facility, and equipment issues as well as staff behavior. The next chapter provides an in-depth discussion of the systems issues, both enabling and sustaining, because their alignment with the desired culture either signals the seriousness of leadership to ensure safety or belies the announced intention to minimize hazards for employees and patients.

Are your leaders aligned on the necessity of a proactive approach to safety? Have they identified the critical upstream behaviors and practices that create or mitigate hazard? Do they track the frequency of these practices? Are positive feedback mechanisms in place to induce increases in the frequency of critical behaviors?

Establishing goals

What should be our patient and employee safety goal? In all industries, leaders set tough targets to challenge the organization and to improve safety performance in the same way they set other operational targets. For example, DuPont is well known for striving to achieve zero workplace injuries and illnesses based on the fundamental belief that "all injuries are preventable." Alcoa, under the leadership of Paul O'Neill, set stringent goals for employee safety and reduced its lost-time incident rate from 1.86 injuries (per 200,000 hours worked) to 0.22. The focus on safety continued even after O'Neill left Alcoa; by 2002 the injury rate had dropped to 0.12.[9] Some healthcare organizations are seeing similar results, relentlessly driving hospital-acquired infection rates, for example, to levels previously thought to be "unachievable."

Atul Gawande describes a simple five-point checklist developed by critical-care specialist Peter Pronovost that plots out the steps doctors need to take when putting a line in a patient.[10] Astonishingly, the checklist reduced the 10-day line infection rate among ICU patients at Johns Hopkins Hospital from 11 percent to zero. At just this one hospital, the checklist prevented an estimated 43 infections and eight deaths and saved the hospital an estimated $2 million. When the checklist was rolled out to hospital ICUs in Michigan, Gawande reports, "Michigan's infection rates fell so low that its average ICU outperformed 90% of ICUs nationwide," saving "an estimated $175 million in costs and more than 1,500 lives... all because of a stupid little checklist."[11]

---

9 Sandy Smith, "America's Safest Companies" (October 21, 2002). Viewed at www.occupationalhazards.com.

10 Atul Gawande, "Annals of Medicine: The Checklist," *New Yorker*, December 10, 2007; pp. 86–95. "Doctors are supposed to (1) wash their hands with soap, (2) clean the patient's skin with chlorhexidine antiseptic, (3) put sterile drapes over the entire patient, (4) wear a sterile mask, hat, gown, and gloves, and (5) put a sterile dressing over the catheter site once the line is in."

11 Ibid.

Even the largest, most complex, and most tradition-bound organizations are capable of order-of-magnitude changes in patient safety performance. In addition to ensuring that a safety goal is set, a board member should feel free to ask what benchmarking has been done in establishing a safety goal, what an improvement versus benchmark organizations would mean to patients in their hospital, what is blocking its accomplishment, and when a new level of accomplishment can be achieved and sustained.

What are your organization's discrete metrics and the explicit goals for safety as a strategic priority? Are leaders aligned on this goal? Do they know their obligations in relation to it? Does the organization realistically believe that all adverse events and injuries are preventable? If not, what obstacles block that belief and expectation?

Once leaders are aligned on the necessity of safety as an organization-wide strategic objective and a core value—and have determined what the alignment means vis-à-vis other strategic objectives—they develop a clear picture of what they expect the organization to look like once it successfully enacts safety as a strategic priority. What will be observably different?

Step 2: Develop a patient safety vision

Leaders often think of an organizational vision statement as something they need to provide employees, and employees often think of the vision statement provided by leaders as management gibberish that has little to do with them and their work. At first, developing a vision statement benefits leaders more than employees. Sharing the safety vision enhances its value throughout the organization.

Whereas strategy is the plan by which leaders intend to accomplish the organization's final objectives, vision contains their ideas about what the organization will look like when the strategic objectives are successfully achieved. Vision is thus a concrete picture of the desired future state. If a company's product strategy involves superior customer satisfaction, the leadership's vision statement may include gathering customer information faster than any competitor in the industry to adapt the product to specific customer wants and needs.

Jim Collins, author of *Good to Great*, says that a vision statement incorporates both the "core ideology" (the creative sum of core values and core purpose) and the "envisioned future." The core ideology specifies what must be preserved in the face of any change, and the envisioned future specifies what will change. The envisioned future comprises "a 10- to 30-year audacious goal... and vivid descriptions of what it will be like when the organization achieves" that audacious goal.[12]

Such a long time frame often strikes the leadership of a healthcare organization as illusory and provokes unnecessary resistance. We sometimes find it useful to remove the long time frame and give expression to a shorter one. For example, one client had trouble seeing the possibility of getting to "zero incidents within 10 years," but discovered that the key participants in the working interface were open to committing repeatedly to "another accident-free day" after a successful year of accident-free operations.

Formulating your strategic vision of safety in terms of an extended time frame, however, provides an important advantage: it incorporates a commitment to sustainability. This commitment builds a degree of protection against the vision's being mistaken as the latest program of the month. Significant safety improvement invariably requires culture change, and successful culture change requires that employees and staff take the new leadership direction seriously. Generating serious support requires significant commitment from leaders. The consequences of a lack of commitment are readily apparent. Employees and staff cynically regard any flavor-of-the-month, uncommitted change effort as worse than no effort at all.

Between Collins's 30-year time frame and the potential exoneration inherent in no time frame, we find it useful to ask the leadership team to look five to 10 years ahead and then back to today as a way of creating a roadmap to the desired future. If you are planning a trip, you need to know where you're headed. This enables you to set direction, specify meaningful milestones, and formulate criteria that let you know when you have arrived. If several people are involved in planning the trip, the first thing they need to do is agree on the destination.

---

12 Jim Collins, *Good to Great: Why Some Companies Make the Leap... and Others Don't* (New York: HarperBusiness, 2001). Viewed at www.jimcollins.com.

As Collins explains, the two components of core ideology and envisioned future induce a creative and dynamic tension between preservation and change. This tension answers the "why?" of a vision statement by acknowledging the difficulty of holding constant the core values while laying out the roadmap to the new destination.

Achieving an organization's strategic goals is a venture into the future. Navigating this venture demands clarity and the alignment of everyone involved. The process entails finding common ground, identifying areas of conflict, and working out differences. Although occasionally arduous, the process yields great rewards.

The rewards of a safety vision

A clear patient safety vision helps leaders:

- Stay focused on achieving the organization's most important objectives
- Avoid diversions
- Resist forces pulling in other directions
- Prevent improvement efforts from becoming watered down and trivialized
- Avoid mistaking a lesser good for a greater one
- Resolve disagreements and overcome narrow self-interest in order to remain on task
- Sustain their shared commitment to the organization's fundamental purpose and core values

When leaders, in turn, share their vision with their entire team, they stand to gain additional value. The picture of the future and the values on which the picture is based give employees and staff guidance for their decision making and behavior. An organization cannot create specific rules covering every situation. In the complex world of healthcare service delivery, the organization has to rely on individuals making independent decisions in the midst of unexpected and unforeseen situations. Variability in these decisions and behaviors is greatly reduced when individuals clearly understand the organization's vision and values. A leadership vision becomes a vehicle for uniting the organization in a common purpose and for facilitating cooperation. As such, it is an important tool for culture change.

A patient safety vision serves employees by:

- Heightening awareness of the individual team member's impact on the organization
- Directing the day-to-day work toward the most important goals
- Creating openings for questioning cultural assumptions and testing the new ideas that culture change requires
- Inspiring employee motivation, clarifying expectations, and raising the bar
- Giving them the strength to endure the difficulties often associated with culture change

Despite these benefits, the payoff to the individual leader may be long term or elusive. Unless leaders become involved out of their own personal ethics—because it is the right thing to do—they are unlikely to persist through the difficulties that culture change entails and are unlikely to be seen as authentic sponsors of the initiative by the teams they oversee and lead.

A safety vision grounded in ethics

The most effective healthcare safety vision is motivated by the leaders' strong, sincere sense of ethical responsibility for the safety of patients and employees. U.S. healthcare causes somewhere in the neighborhood of 50,000 to 100,000 *preventable* patient deaths a year.[13] U.S. industry long ago took up the challenge of safety, and the number of fatalities that occur in industry is about 5,700 per year, across all manufacturing industries, government agencies, construction, and transportation.[14] Many in healthcare share a sense of shame. How can we—of all industries!—lag so far behind? What forces have brought about this deplorable situation?

Ethical leadership is the fertile soil in which successful changes in organizational culture flourish. It provides a basis for enlisting

13 Institute of Medicine, *To Err Is Human: Building a Safer Health System*, eds. Linda T. Kohn et al. (Washington, DC: National Academy Press, 1999). Viewed at www.nap.edu.

14 U.S. Department of Labor, Bureau of Labor Statistics, *National Census of Fatal Occupational Injuries in 2006*. Viewed at www.bls.gov.

and motivating others to set aside their narrow self-interest and cooperate to build a better future. The temptation to suppress our ethical impulses often stems from our mistaking an extrinsic good for something of intrinsic value. For example, in healthcare there is intense competition for important but scarce extrinsic goods. Money, power, career, and status sometimes conflict with our ethical motivations, as they do elsewhere. Therefore, ensuring patient safety means bridging the divides of self-interest. It may also require sticking one's neck out: authority relationships are strong in medicine. Authority can make people hold their tongues and obey leaders to the exclusion of their better impulses and better judgment. Moreover, burnout is a constant threat in healthcare, robbing providers of the energy and involvement they need to even care whether something is right or wrong.

Visible examples of ethics at the top gives team members the support they need to take positive actions such as these:

- A surgeon voluntarily takes time away from his busy practice for a "nonmedical" issue like patient safety.
- A physician takes the initiative to confront weaknesses in her profession's system of self-regulation and self-policing.
- An administrator puts patient safety on a proper footing vis-à-vis financial considerations.
- A department head sets aside his impulse to protect his organizational or professional turf.
- A nurse raises her concerns with the physician instead of simply carrying out his orders, when she believes that doing so would harm the patient.

It is natural to feel that we will automatically do the right thing in situations like these, but the right action often poses more difficulties than we anticipate. For example, Philip Zimbardo[15] points out how authority relationships can create unquestioning obedience and evil behavior. He cites a well-known study

15 Philip Zimbardo, *The Lucifer Effect: Understanding How Good People Turn Evil* (New York: Random House, 2007), p. 277.

from the 1960s involving a group of nurses. Experimenters had an unknown M.D. phone in an order calling for the injection of a (placebo) medication from a vial labeled in such a way as to make it obvious to the nurse taking the call that the dose ordered was potentially lethal. In the experiment, the physician called 22 nurses individually. All but one of the 22 carried out the order without question.[16] In a more recent study cited by Zimbardo, 46% of nurses surveyed said that they had carried out orders they believed to be harmful.[17]

Ethical leadership asks, What kind of person am I? Do I want to be the sort of person who does the right thing in this situation? What must I do to be fully responsible for my behavior?

Ethical leadership proceeds from the leader's commitment to doing the right thing. It asks each person, when confronted with safety-relevant situations such as those just mentioned, to inquire of himself or herself: *What kind of person am I? Do I want to be the sort of person who does the right thing in this situation? What must I do to be fully responsible for my behavior?*[18] Moreover, ethical leadership actively supports the dialogue necessary to resolve difficult ethical issues appropriately. Dialogue doesn't mean discussion and debate. It means the sort of collective revelation and suspension of bias described by physicist David Bohm:

> The object of a dialogue is not to analyze things, or to win an argument, or to exchange opinions. Rather, it is to suspend your opin-

16 C. K. Hofling et al., "An Experimental Study in Nurse-Physician Relationships," *Journal of Nervous and Mental Disease,* 143 (1966): 171–180.

17 Annamarie Krackow and Thomas Blass, "When Nurses Obey or Defy Inappropriate Physician Orders: Attributional Differences," *Journal of Social Behavior and Personality,* 10 (1995): pp. 585–594.

18 Zimbardo, *Lucifer Effect,* p. 275.

> ions and... to listen to everybody's opinions, to suspend them, and to see what all that means... And if we can see them all, we may then move more creatively in a different direction.[19]

Ethical leadership empowers others to do the right thing in tough situations. In such ethical soil, the effective vision takes root as the practice of proactively minimizing or eliminating hazards in the working interface and expanding the range of incidents that are preventable. Vision flowers by continually improving all the elements in the Blueprint for Healthcare Safety Excellence (see Figure 2–1 in chapter 2):

- Leadership
- Organizational culture
- Healthcare safety–enabling elements
    - Hazard recognition and mitigation
    - Skills, knowledge, and training
    - Regulations and accreditations
    - Policies and procedures
    - Patient safety improvement mechanisms
- Organizational sustaining systems
    - Organizational structure
    - Selection, development, and retention
    - Alignment
    - Performance management
    - Rewards and recognition
    - Employee engagement systems
    - Management systems
- Working interface

To build the kind of culture they are aiming toward, great safety leaders take the initiative to share their vision for healthcare safety and the ethical foundation upon which it rests. They

---

19 David Bohm, *On Dialogue* (Ojai, CA: David Bohm Seminars, 1990), p. 14.

do so frequently and widely—with all employees and treatment team members. They also enlist additional leaders as champions of the vision.

Living and sharing the safety vision

To be effective in sharing their vision, leaders authentically describe, explain, and discuss the vision in their own words, not to mention act in accordance with it. It helps to embody the vision in a narrative or story describing how the leaders came to have this vision, how things will be different and better, and why the vision is both necessary and desirable.

How will people act differently from the way they do now? In terms of the dimensions of the Organizational Culture Diagnostic Instrument (OCDI), for example, the quality and frequency of both upward communication and approaching others will measurably improve. How will priorities be different? Safety itself will move from a priority to a value, and the gathering of information on near misses and recurring hazards will command new importance and attention.

Which decisions will be made differently? Critical safety behaviors will be identified, codified, disseminated, and observed more commonly at the leadership level and in the working interface; purchasing decisions will include the routine evaluation of safety implications.

How will each element in the Blueprint for Healthcare Safety Excellence be different? Safety considerations and means of compensating for cognitive bias will become routine parts of decision making for staff and employees at all levels. Every safety-enabling and sustaining system will undergo examination and adaptation to ensure a faster and more effective response to the new organizational value for safety. (The next chapter discusses these systems in detail as a part of the implementation of the desired culture change.) These systems send loud signals about the intentions of leadership. Maintaining the status quo means no real change is expected or desired.

How do the governing body and the senior leadership monitor organizational culture? On behalf of the board and the CEO, an OCDI is administered regularly (e.g., annually) to give the top of the house measures of the improvement in observable behaviors known to be predictive of safety improvements. As mentioned earlier, we use only a validated diagnostic instrument; a survey may provide interesting anecdotal evidence but not the comparative grounding that is predictably related to safety outcomes.

Sometimes leaders are uncomfortable talking about their vision and its ethical foundation. They worry they will sound too idealistic or preachy. To become more comfortable and persuasive when talking about these things, each leader composes and rehearses a three-minute, conversational speech to explain the leadership vision and the ethics behind it in his or her own words. Leaders use their own illustrations and examples of the organization's core values and how things will be different, including behaviors and practices that are familiar and important to their direct reports.

The leaders share their talks with each other to calibrate their alignment, freely borrowing each other's most emblematic examples. They also talk frequently with their direct reports about the vision in order to become sensitized to the cultural attributes that need to change. As they encounter instances of these attributes, leaders use them as opportunities to discuss the vision. Discussing doesn't mean lecturing or preaching. It means having a two-way conversation, considering new ideas, talking, and thinking together about what the vision entails for each participant in the conversation, seizing the opportunity to challenge and inspire each other about patient and employee safety.

Such conversations mobilize people's energy for a more desirable future. By enlisting many people in the change effort, the conversations give the culture greater flexibility and numerous opportunities for "change experiments." This conversational culture, in turn, produces better solutions concerning how to accomplish the desired future.

Leaders formulate goals and jointly create projects with their reports that are consistent with the vision and foster its implementation. Finally, each leader enacts the vision in her behavior and decision making.

One way a leader can strengthen his ability to walk the talk is to identify situations in which he fears he will have difficulty and make a practice of asking for regular 360-degree feedback on his performance in these areas. In addition to benefiting the leader, this practice fosters the open communication necessary to a strong safety climate and outstanding performance. In addition to each leader's taking individual action in support of the vision, leaders *as a group* need to systematically evaluate the organization's strengths and identify where the organization falls short of its safety vision.

Step 3: Perform a current state analysis

Making safety a strategic objective usually entails considerable change in a healthcare organization, both structurally and culturally. In an ideal world, the outcome looks like this:

- The board of directors regards the governance of strategic objectives as its responsibility and expects the organization's executives to deliver measurable progress against these objectives.
- Safety holds a prominent position among the organization's strategic objectives, and the board expects results. The board receives reports on meaningful and proactive measures of safety and the progress of safety initiatives.
- The chief safety officer customarily reports results to the CEO. Safety projects routinely appear in the business plan and budget.
- The vice president of safety commands the necessary resources as well as the support of the CEO and other leaders as she acts across organizational boundaries to improve safety.
- Safety performance is central in decisions about hiring, compensation, and advancement.
- Safety plays a prominent role in all operational decisions.
- All employees understand that safety is part of their job. They have the resources to accomplish it, know how to do it, and actively work at it.

For most organizations, reaching this desired state requires many changes. Leaders need to pinpoint the specific changes required for the organization's transformation and then order them according to their relative urgency and potential net benefit. In the Patient Safety Academy leaders gather data in order to understand and fully appreciate their organization's current safety performance and the factors that influence it. Each element in the Blueprint for Healthcare Safety Excellence is evaluated and summarized for presentation to the leadership team and constituencies. The presentations make explicit the extent of the organization's exposure as well as the adequacy of its safety metrics.

Gathering cultural data

Information about the current state of the culture gives leaders many of their most important improvement targets. Because distinct subcultures often exist in different facilities and their subunits, the diagnostic data should be obtained in a sufficiently robust way that allows the information to be disaggregated and analyzed by location and subunit.

Leaders need information about all nine dimensions of the OCDI, explained in chapter 3 and recapped in Table 8–2 for ease of reference.

TABLE 8-2. ORGANIZATIONAL CULTURE DIAGNOSTIC INSTRUMENT (OCDI).

ORGANIZATIONAL DIMENSIONS	TEAM DIMENSIONS	SAFETY DIMENSIONS
• Procedural justice • Leader-member exchange • Leadership credibility • Perceived organizational support	• Teamwork • Workgroup relations	• Organizational value for safety • Upward communication • Approaching others

Of the nine dimensions of the OCDI, the first six address the organizational and team "glue" that fosters loyalty and good citizenship. These dimensions address how workers and teams at all levels relate to one another and function effectively. They underlie the excellence of performance and reliably predict the tendency toward ethical thinking and practices as an organization. The remaining three safety dimensions quantify the organization's commitment to safety relative to that of other complex organizations. The continual measurement of these elements of culture and climate provides the senior leadership and the board with the information they need to fulfill their responsibility of ensuring the right tone not only at the top but also in the working interface.

How does your organization compare with others on these dimensions, and what are the implications?[20] Where are the greatest vulnerabilities? Which areas will return the greatest performance improvement for effort expended? Who are the key players for each area of opportunity?

Although the OCDI provides the kind of information needed to identify target areas, leaders also need to be aware of any specific cultural red flags that foretell serious safety or ethical lapses. These red flags include:

- Exceptionally low scores on one or more OCDI scales (low scores on upward communication are especially troublesome)
- Unusually high levels of competition for financial rewards due to either explicit incentives or implicit aspects of performance management
- Weak or thinly supported line supervision
- Financial insecurity in the organization generally or in specific services[21]
- Exceptionally difficult time pressures (e.g., in the emergency service or cardiac ICU)
- Situations that chronically create fatigue (e.g., sleep deprivation during internship and residency training has been shown to significantly increase error rates)
- Slow or poor quality record-keeping

---

20 Again, a survey may provide useful insights and stories but may not allow strategic comparability at this stage of the change effort.

21 Or large differences in dedicated endowments among departments or programs in a major academic medical center.

- Staff cynicism or attitudes of entitlement[22]
- Any prevalent dehumanizing or alienating defense mechanisms among staff[23]

Interviews and focus groups with leaders, managers, and treatment team members are conducted to cross-validate the OCDI findings and gather in-depth knowledge about specific issues.

Gathering leadership data

Leaders lead within a unique cultural context, and their behavior—especially their leadership style and best practices—shapes that context. Each leader needs to find out, *How does my behavior contribute to our cultural problems? How do I need to behave differently to bring about improvements in our organizational culture?* The Leadership Diagnostic Instrument (LDI), introduced in chapter 4, gives the leader the information she needs to think productively about this question by providing comprehensive feedback about her relationships with important others in the organization.

Gathering hazard data

To understand the safety situation in the working interface, leaders ask four questions:

1. What are the major categories of hazard in the working interface?
2. What is the level of hazard manifest there and how is it changing over time?
3. How engaged in hazard reduction are those who work there?
4. How effectively are hazards being reduced and eliminated?

Several methods are available for understanding the hazard categories in the working interface. Techniques drawn from Six Sigma, Lean Systems, and behavior-based methodology can all be

22 See Philip Zimbardo, *The Lucifer Effect: Understanding How Good People Turn Evil* (New York: Random House, 2007) and *Nature* Podcast on the dark side of science: research misconduct, January 18, 2007, based on the article by Jim Giles, "Breeding cheats," *Nature*, 445 (January 2007): pp. 242–243.

23 Zimbardo, *Lucifer Effect*.

used. The hazard level is measured directly using behavior-based methods once safety-critical behaviors and practices have been identified.

Gathering team data

The level of team engagement in hazard reduction is measured by means of the three safety-specific dimensions of the OCDI, augmented by interviews with employees and professional staff to determine:

- Do those who work in the working interface know which of their behaviors and practices are critical to safety?
- Do they understand and recognize the practices and situations that most frequently increase the level of hazard where they work?
- Do they have the resources they need to perform safely?
- Is safety a positive issue for them, or is it fraught with conflict, criticism, and fear?
- Do they understand that safety is an ethical imperative?
- Do they feel they have meaningful opportunities to make a difference?
- In what ways are they engaged?

With adequate information about the current state in hand, leaders are ready to take the next step: developing a high-level plan to close the gap between the current state and the vision.

Step 4: Develop a high-level intervention plan for phase II

After aligning on safety as a strategic objective, developing a patient safety vision, and analyzing the current state, leaders develop a high-level safety improvement plan—or they put in motion a mechanism by which a plan will be developed by a team from within the senior leadership group. The intervention plan identifies areas where the current state diverges materially from the leadership vision. It encompasses change objectives and addresses how needed changes will be set in motion, resourced, and monitored.

The plan explicitly answers these questions:

- What exactly is the objective of phase II? What is its scope? The objective and scope, which will vary depending on the current state and the leaders' vision, can best be spelled out in the terms of the elements in the Blueprint for Healthcare Safety Excellence. In most situations, there will be improvement objectives and intervention scopes for each element.
- What specific outcome measures will be tracked to determine if the plan is meeting its objectives? What are the goals for these outcome measures over what period?
- Which healthcare safety–enabling elements and organizational sustaining systems (for convenience, referred to collectively as "systems") are most likely to impede the achievement of the new vision? Who is responsible for analyzing these systems and determining the level of fix needed in order to align each system with the new strategy and vision? This question sets the stage for phase II, detailed in the next chapter.
- By what process will phase II be carried out? Who will be responsible for doing what and by when? How will key actors be identified, selected, enlisted, and empowered?[24]

---

24 Leaders can identify and recruit key players using an assessment center methodology that identifies the specific knowledge, skills, and abilities required for success at a given job, and then tests candidates against these criteria by having them actually perform tasks that are representative of those they would perform if accepted for the position. This testing often entails role-playing scenarios that call upon the relevant knowledge, skills, and abilities. Not only does this methodology select the best candidates for the job but it also identifies each candidate's specific gaps. It tells you what training or development a candidate will require to succeed at the job. What does not work is assigning key roles uncritically based on the candidate's current position, willingness to take on a role, seniority, or availability.

Healthcare organizations often charter an implementation team to do much of the phase II work. Whatever the case, reporting relationships for the people doing this work should be clearly specified, and progress should be monitored, measured, and acknowledged by leadership.

Additional questions to be answered in the plan:

- Will the intervention be carried out at the location level or from the corporate group?
- How will resources needed for phase II be identified and allocated? Oftentimes, the budget to get things started comes out of the CEO's discretionary fund or a special board allocation, at least until safety can be budgeted as a strategic objective in the next planning cycle.
- What will not be delegated, and who will be responsible for those elements? Again, the responsibility for defining the direction and changing the culture cannot be delegated. In addition, leaders must remain involved to provide focus, oversight, resources, and recognition. Leaders' continued participation ensures that the organization stays the course.
- To whom, how, and when will the initiative's work product be communicated—and by whom? When the initiative commands strategic priority, interim results are regularly reported to a review body of senior leaders, board members, or both. In addition, telling the story of the project (or of the senior planning retreat and its accomplishments) can be an effective way for leaders to initiate the process of moving the organization toward its vision.

The high-level plan should not lay out specific solutions in great detail because working out the solutions and then implementing them can provide important engagement opportunities for everyone during the planning and implementation stages. Involving leaders, managers, treatment team members, and other employees in these activities produces higher quality results.

Taking advantage of such opportunities for engagement fosters the desired culture change. It is important, nonetheless, that leaders provide visible and active support as well as direction, perhaps even an explicit charter, for those who will be doing the phase II design and implementation work.

Here is an example of one charter that outlines the roles of the safety team and the larger organization:

> The Safety Design and Implementation Team is chartered by the Medical Center to plan, develop, implement, and supervise the installation and initiation of our Leading with Safety culture improvement process under the direction and oversight of the Safety Senior Leadership Team. The results of the team's work will be reported to the Medical Center's senior leadership and to the risk management committee of the board of directors/trustees.
>
> The Safety Design and Implementation Team (of no more than 12 members) will be representative of the constituencies, including professional groups, management, and other employees from a sample of departments and services. The Medical Center will provide the Safety Design and Implementation Team with the resources and authority it needs to accomplish its mission.
>
> A designated facilitator will assist the Safety Design and Implementation Team. A designated member of senior leadership will sponsor and advise the team and will work with the facilitator to manage resources and track the progress of plan design and implementation. The facilitator and leadership sponsor will report monthly at the Safety Senior Leadership Team meeting.
>
> The Safety Design and Implementation Team will recruit and train others to: (1) carry out the process on a continuous basis and (2) establish and maintain a strong organizational culture and safety climate, plus a high standard of safety for all Medical Center patients and employees.

In this chapter we focused on leadership's role in the culture change process. During phase I, in the Patient Safety Academy, leadership aligned on safety as a strategic priority, developed its patient safety vision, reviewed the organization's current state, and developed a high-level plan to achieve that vision. The next chapter roughs out phase II, implementation of that plan. We say "roughs out" because no two implementations follow identical paths.

CHAPTER NINE

LAUNCHING CULTURE CHANGE FOR *PATIENT AND EMPLOYEE SAFETY*

LAUNCHING CULTURE CHANGE FOR PATIENT AND EMPLOYEE SAFETY

Phase I of the Leading with Safety process aligned leaders on the strategic role and value of safety, elicited their vision, highlighted the gap between the vision and the current state, and developed a high-level plan for closing the gap. Phase II fleshes out that plan and drives safety throughout the organization to the behavior level. Phase II engages personnel, improves systems, and maintains a focus at the level of safety-critical behavior. It also generates and uses new safety data to initiate continuous improvement and sustain the improvement process.

Phase II: Achieving safety throughout the organization

Whereas leaders performed all the phase I work, much (but not all) of the phase II work is delegated. Leadership's role in phase II is to guide change in organizational culture, plus:

- Provide direction, focus, oversight, resources, and recognition.
- Realign systems, both healthcare safety–enabling and organizational sustaining.
- Model new, safety-critical behaviors and best practices.
- Personally ensure that the organization stays the course.

- Participate with everyone else as subjects of the intervention.
- Engage others to detail the implementation plan.

An ambitious phase II plan may take several years to achieve its objectives, but measurable progress should be expected in the first year. Phase II employs safety as a rallying point around which to create deep and far-reaching culture change; it leads to continuous safety improvement. At the less ambitious end, phase II may aim simply to create a leadership umbrella under which the many disparate safety improvement initiatives already under way in the organization are incorporated, integrated, and elevated in status. Or phase II may be used to refocus organizational awareness on patient safety in an already high-functioning institution. In this case, phase II may consist of patient safety workshops for employees and professional staff. (Even in a more ambitious plan, such workshops have stand-alone value and may serve as a kickoff for other, more aggressive interventions.)

Chronic curiosity is the way to achieve patient and employee safety.

To recap the phase II steps (see Figure 8–1):

- Step 5: Engage the organization in the Leading with Safety process.
- Step 6: Realign systems, both healthcare safety–enabling and organizational sustaining.
- Step 7: Establish a system for behavior observation, feedback, and problem solving.
- Step 8: Sustain the Leading with Safety process to continually improve the organizational culture and patient safety.

After we discuss each of the phase II steps we will present a case history that touches on some highpoints of the Leading with Safety process and makes use of the tools we have presented in this book.

Step 5: Engage the organization in the Leading with Safety process

Organizations that achieve high levels of safety performance find ways to engage staff from all disciplines and at all levels. True engagement requires well-functioning processes and systems. It also creates personal commitment and personal accountability, both of which are critical in improving patient safety and creating a performance-oriented culture.

Instilling dialogue and transparency throughout the organization

Dialogue is essential to effective collaboration about safety. The first value in the Mayo Clinic's model of patient care reads: "Collegial, cooperative teamwork with true multispecialty integration." As new patients at Mayo soon discover, any uncertainty about their diagnosis or treatment provokes multispecialty dialogue and collaboration as well as intensive conversations with the patient.

The principles of dialogue and transparency should apply to the whole organization, from top to bottom. Wachter and Shojania describe "a new culture in which residents speak openly about their mistakes and help devise solutions to the problems that caused them."[1] We wonder why the same principle can't apply to board members and the executive leadership as well. Such openness would promote more frequent and effective upward communication throughout the organization.

James Reason says that the price of patient safety is "chronic unease."[2] We prefer to think that chronic curiosity is the way to achieve safety for both employees and patients. Constantly asking what we know and don't know, pursuing what has changed and why, inquiring about what went before and after an undesirable but preventable outcome, and exploring which leadership behaviors may have signaled the very behavior that occasioned the adverse event—these questions all bring continual improvement.

---

1 Robert M. Wachter and Kaveh G. Shojania, *Internal Bleeding: The Truth Behind America's Epidemic of Medical Mistakes* (New York: Rugged Land, 2005), p. 359.

2 Ibid., p. 361.

The board can add a single agenda item: extended conversations in response to the question, "What do we know and what don't we know about how we're doing with patient and employee safety?" That conversation should prompt another one: "And what should we do about it?" People sometimes have the answers but are afraid to speak up. When senior leaders present self-critical reviews of their own safety-supporting behaviors, this creates a safe space for the continual upward reporting of near misses and mistakes.

Tell patients where the safety conditions are uncertain.

The transparency needs to extend to patients as well. Jerome Groopman supports Jay Katz's contention that the uncertainties of healthcare should be acknowledged to individual patients.[3] We would extend the principle to the organizational level and expand it to safety: Acknowledge to all patients where the safety conditions are uncertain so they can compensate for any condition that might unnecessarily expose them to hazard. That is, teach patients how to use your organization well. They can help mitigate hazards, e.g., by noting whether a provider entering the examination room washes her hands before examining them—and by speaking up if the provider doesn't do it.

Selecting champions

Who should drive safety at each level, within the various locations or system-wide? This question begs for both a committee answer and a chain of command answer. But neither is exclusively the driver. Safety requires an ongoing exchange of information among peers and a two-way, vertical flow of information to reveal the full iceberg of hazards.

3 Jerome Groopman, *How Doctors Think* (Boston: Houghton Mifflin Co., 2007), p. 152: "Katz argues... the greater burden is 'the obligation to keep these uncertainties in mind and acknowledge them to patients.'"

Because different types of expertise are required, patient safety management functions are distributed to specialized groups of professionals within healthcare organizations. Thus the real answer to "Who is driving patient safety?" may differ from one part of the organization to another. But the chain of command governing safety is only as strong as its weakest link. Every level and function of the organization—from the boardroom to the patient care units—has an active role in the organizational mechanisms for minimizing exposure to hazard. What matters most is that every level and function takes steps that can be observed, verified, reported, and modified as the organization learns how to optimize its own patient safety performance. The visibility and malleability of these steps represent the first level of transparency in patient safety.

Engaging staff

Engaging staff means more than putting up posters and holding safety fairs. Almost without exception, staff members have a deep commitment to the safety of patients, and they are open to becoming engaged. But effectively engaging them requires leaders who place a high value on patient safety and who express the value consistently in their actions, words, and decisions.

Staff members become engaged when given opportunities to become actively involved—both formally and informally. Strong safety leaders create the expectation of such involvement and provide the needed resources, opportunities, and support. Staff can become formally engaged in hazard reduction at many levels. They should be encouraged to present at a senior leadership meeting, in a safety meeting, or during safety rounds; to sign on to a safety problem-solving team; and to take on a larger commitment such as sitting on the chief safety officer's safety design and implementation team. Whatever form staff engagement takes, the staff's commitment to patient and employee safety needs to be fostered and recognized by the organization, especially by the team members' superiors.

Engaging supervisors

Supervisors build safety on the foundation of the relationships they establish with their direct reports. To support employee and patient safety and reduce hazards in the working interface, supervisors should:

- Avoid the language of blame.
- Reinforce even small efforts at upward communication about safety.
- Expect, encourage, recognize, and reinforce the safety activities of direct reports.
- Provide the time and resources needed by reports who are active in other safety roles (and backfill, if necessary, to avoid a backlash against employees who are "away" from their primary jobs).
- Respond to systems problems and follow through on safety commitments in a timely manner.
- Provide frequent feedback to reports on the status of hazard reduction efforts.
- Explicitly make safety considerations a part of all operational and clinical decisions.
- Communicate effectively and candidly, both upward and to direct reports, on behalf of safety.
- Take on safety-specific roles (such as serving on a safety committee or problem-solving team) as necessary.

Engaging managers

To engage managers in the Leading with Safety process, accountability and performance mechanisms are put in place that recognize managers for their implementation achievements and tangible successes in mitigating hazards. The role of managers is to:

- Receive, review, and use safety data.
- Monitor and reinforce supervisor safety performance.
- Positively reinforce the upward communication of exposures and near misses.
- Give safety considerations a prominent and explicit place in clinical, operational, and staff selection decisions.

Managers also play a crucial role in identifying, diagnosing, and participating with senior leadership in solving system problems. Improving systems provides a wide range of opportunities for culture-building engagement, and managers are in a position to identify employees and staff who should be involved. Involving concerned employees in system improvement projects produces better results both in terms of specific solutions and the ease with which solutions are embraced by users. An appropriate senior leader should oversee system improvement projects to ensure integration with other systems and to optimize potential synergies—and, most important, to ensure that the process by which the project is carried out maximizes the opportunity to influence the culture positively.

Step 6: Realign systems, both enabling and sustaining

Managers and leaders need to understand the degree to which healthcare safety–enabling elements and organizational sustaining systems (for convenience, referred to collectively as "systems") are in tune with safety as a strategic priority. Are any systems missing? Are all needed systems present but some functioning at a low level? Are they present and functioning well but not tuned to achieving safety as a strategic priority? Do recurring hazards and accidents stem more from the lack of guidance or from weak implementation?

This task can be seen in part as a signaling issue. Once the change vision and strategy are in place and well understood, the organization from top to bottom will interpret the authenticity of the change initiative on the basis of the degree to which enabling elements and sustaining systems reinforce the leadership's expressed intention.

Healthcare safety-enabling elements

As you may recall from Figure 2–1, healthcare safety–enabling elements include such things as hazard recognition and mitigation; skills, knowledge, and training; regulations and accreditations; policies and procedures; and patient safety improvement mechanisms. There are two classes of enabling elements: those that provide protection against a limited range of specific hazards, and those that broadly mitigate hazards. A hand hygiene policy and deployment of disinfectant dispensers exemplify the first. Hazard recognition training, change management efforts, and the gathering of safety data illustrate the second, as does a set of safety procedures targeting high-risk activities. Healthcare organizations usually need both types of safety-enabling elements.

To realign the safety-enabling elements, leaders first need to understand their organization's specific safety vulnerabilities. The phase I current state analysis (step 3) should evaluate the adequacy of enabling elements for the identification and mitigation of hazards as well as the elements' user-friendliness and efficiency, plus the extent to which they are actually used. In addition to the administrative review of enabling elements, a useful source of information is the people who face safety problems regularly in the working interface and who rely on these systems for hazard mitigation.

Safety performance and organizational sustainability are closely linked.

Many enabling elements are static—i.e., they reside in written documents and procedures but are not used regularly as part of the day-to-day activity of the working interface. For example, staff frequently fail to wash their hands despite the written policy and convenient opportunities to do so, just as they fail to report incidents or avail themselves of safety suggestion opportunities.

Because of this behavior failure, high-functioning organizations institute positive mechanisms of behavior observation and feedback to ensure that systems are actually used and critical behaviors are performed consistently and reliably. (We'll talk more about behavior observation and feedback in step 7.)

Safety data

What safety data are available? To whom? How and how well are they used? Is the board of directors or trustees routinely informed about adverse events, issues, initiatives, root causes, and progress in mitigating hazards upstream of major events? Years of conducting safety interventions and studying this issue have taught us that safety performance and the organization's sustainability are closely linked. Linking safety with organizational performance requires valid patient safety metrics, a clear goal, an awareness of early indicators, an understanding of the empirical relationship between patient safety and other performance metrics, plus mechanisms that build and sustain safety climate and reach to the behavior level. An organization unable to link safety with other aspects of performance undermines its leaders' capacity to sustain safety as a strategic priority.

Leaders need to understand these relationships for their own organizations. Board members deserve to know and have a duty to understand these connections in a rigorous and validated way so as to optimize the board's value contribution to the organization's stakeholders.

In the working interface, a dearth of candid incident reporting and rigorous, proactive safety data can be debilitating. In the absence of such data, hazards will not be identified nor adequate solutions developed until too late, i.e., after incidents occur. Many organizations develop the capacity to identify hazards, analyze their root causes, and devise intervention plans but then drop the ball when it comes to implementing and sustaining those plans and following up to make sure the plans have the intended results.

Responsiveness to change

A recurrent theme in healthcare—far less prevalent in other industries—is real-time, trial-and-error problem solving in response to difficulties created by change. It is important for leaders to understand how enabling systems will be maintained in the face of change.

The need to manage change applies not only to exceptional circumstances but also to the many routine change events that occur constantly on any busy healthcare service, events such as patient admissions, discharges, and handoffs; the fluctuating level of demand for emergency or intensive care services; and staffing changes. For example, during Hurricane Katrina, emergency generators located on the ground level of one hospital were inundated by floodwaters. In a non-Katrina-related example at another hospital, a patient died when the medical staff was unable to transport him to intensive care during an emergency because the service elevators had been temporarily decommissioned—one for a repair and another for a remodel—and, in the absence of a coordinated change management process, no one had anticipated and provided a remedy for the problems this decommissioning could cause.[4]

Organizational sustaining systems

Organizational sustaining systems include organizational structure; selection, development, and retention; alignment; performance management; rewards and recognition; employee engagement systems; and management systems—in other words, mechanisms that sustain the positive effects of the safety-enabling elements. To realign systems for greater effectiveness, the organization asks itself:

- Does the selection and chartering of safety leadership confirm the strategic importance of safety for both employees and patients?
- Do selection and promotion criteria explicitly acknowledge the enhanced value of safety to the organization?
- Do measurement systems give prominence to upstream measures of hazard while still keeping the entire organization fully aware of outcomes?
- Can leaders be observed learning critical new behaviors; e.g., how to encourage and welcome the reporting of near misses, and do they capitalize on this information to reduce future hazards?
- Does the safety initiative take immediate steps to fill in the complete picture of safety data from leading to concurrent to post-event data collection, reporting, and trend tracking?

---

4 Wachter and Shojania, *Internal Bleeding*, pp. 169–174.

- Does leadership analyze and report widely in the organization the root causes of exposures, near misses, and incidents?

The realignment of certain key systems—recognition and reinforcement, and how the safety function is organized—is especially important.

Organizing the safety function

Wachter and Shojania offer a good suggestion: "Each hospital should have a Patient Safety Director—and preferably two: one a physician and one a nurse or pharmacist—with the clout, independence, and *resources to measure errors,* conduct routine audits of error-prone areas, coordinate detailed investigations, and follow them up..."[5] If patient safety directors are given the resources to measure errors as well as the upstream, proactive, safety-critical behaviors that mitigate exposure to harm, the safety directors also have the resources to report near misses and errors internally and to alert the entire organization—from the operating room to the boardroom—about these error-producing conditions.

A large organization may need a vice president of safety (or a vice president of quality and safety). The controlling principle, of course, is that the position encompass whatever is required to make safety a strategic priority—in terms of access to senior leadership and the board, ability to use and adapt critical information systems, and the latitude to direct (or redirect) resources as needed to prevent hazards.

Safety's focus is another important issue: Is it devoted primarily to compliance, or is a significant share of its activities and resources devoted to proactively improving performance? When devoted only to compliance, safety loses strategic clout and runs the risk of taking on cost minimization, rather than performance optimization, as its chief goal.

Rewards and recognition

Do sustaining systems for hiring, credentialing, promoting, determining bonuses, and providing recognition drive toward safety as a strategic objective? Do they put safety in conflict with other

---

5 Wachter and Shojania, *Internal Bleeding,* p. 348. Emphasis added.

objectives? We have argued elsewhere that incentives should not be provided directly for safety outcomes.[6] This view stems from our conviction that 1) safety is not a *cause célèbre* deserving special attention, and, as a corollary, 2) safety excellence is valuable in its own right.

In organizations that may operate with a top-to-bottom incentive philosophy (e.g., those that have organization-wide "success sharing"), we urge the use of direct rewards for the behaviors that mitigate or eliminate hazards *well upstream* of any safety incidents. Direct incentives for safety outcomes tend to drive underground the rapid reporting of near misses, de-emphasize the mitigation of hazard, and retard the elimination of underlying causes.

> Direct incentives for safety outcomes tend to drive near-miss reporting underground.

Among our leading industrial clients, the companies setting the benchmark for industry safety often have the highest rates of reported near misses because they do not penalize the reporting of near misses and do not directly reward the reduction of incident rates. Instead, they welcome the information stemming from near misses, quickly digest its implications, and act immediately to reduce the likelihood of repeated exposures to hazard.

Because of improved reporting, high-functioning organizations often find that their incident frequency numbers initially increase in the short term as they establish safety as a core value. In spite of this common short-term effect, industrial clients on average improve more than 40% in the first year.[7]

---

6 See "Transitioning Away from Safety Incentive Programs," in *The Behavior-Based Safety Process: Managing Involvement for an Injury-Free Culture*, by Thomas Krause (New York: John Wiley & Sons, 1995), pp. 317–330.

7 Kristen Bell, Matt O'Connell, Matt Reeder, and Rebecca Nigel, "Predicting and Improving Safety Performance," *Industrial Management*, April 2008.

Step 7: Establish a system for behavior observation, feedback, and problem solving

Healthcare is not unique with respect to the issue of behavioral reliability. Virtually every organization experiences a significant gap between the way things are done and the way they are planned to be done. High-reliability organizations—those that function at a high level and get excellent results—find ways to address these issues. For example, how do we ensure that caregivers practice good hygiene, not usually or frequently but reliably, all the time, across the entire healthcare system? More broadly, how do we ensure that the thousands of behaviors that put patients and employees at risk are systematically reduced in frequency?

The naïve observer may see this task as an impossible challenge. People are human; how can we realistically expect them always to do things the way the procedure requires? Our experience and data[8] have shown that a strong organizational culture and safety climate with good safety-enabling elements and sustaining systems can achieve high levels of behavioral reliability. Doing so is a process, and it has identifiable elements and steps.

The process consists of identifying safety-critical behaviors, systematically measuring the frequency with which they are performed in a safe manner (rather than not performed at all or performed in an at-risk manner), and using this information for performance feedback and as a means to identify barriers to safe performance. This process provides employees and staff with positive, real-time feedback as well as nonpunitive, corrective feedback. Regular measurement also occasions frequent, relevant, brief safety conversations that provide soon, certain, and positive feedback to caregivers and other employees (you'll recall, from the discussion of applied behavior analysis in chapter 6, the pivotal role of positive feedback). The measurement process generates ongoing, proactive data about the level of hazards in

8 Thomas R. Krause, K.J. Seymour, and K.C.M. Sloat, "Long-Term Evaluation of a Behavior-Based Method for Improving Safety Performance: A Meta-Analysis of 73 Interrupted Time Series Replications," *Safety Science*, 32(1999): pp. 1–18, and Thomas R. Krause, *Leading with Safety* (New York: Wiley & Sons, 2005).

the working interface as well as information about barriers to improvement.

The process of identifying hazards in the working interface, measuring the frequency of their occurrence, and recognizing safety barriers and solutions constitutes a continuous improvement cycle for safety that creates a positive safety climate and drives cultural improvement. This process isolates specific behaviors to target for improvement. It also provides a common language for safety and stimulates both horizontal and vertical communication about important safety issues.

Although leaders often anticipate treatment team resistance to behavior measurement, we find that good leadership can readily overcome such resistance. Leaders are often surprised at how effectively an implementation team can champion the measurement process and secure the willing participation of treatment team members and peers. The behavior measurement process contributes in several ways to successful, far-reaching, deep culture change:

- It enacts safety as a strategic value and drives culture change through the lowest levels in the organization.
- It creates opportunities for productive and protected dialogue.
- It generates important proactive safety data for leaders.
- It provides invaluable data for use in safety problem solving.
- It creates many meaningful opportunities for treatment team and employee engagement.
- It directly reduces working interface hazards and improves the level of organizational safety performance.
- It helps teach patients and their families as well as other external caretakers how best to participate in medical error prevention.

As expected, when safety projects are successfully implemented and hazards mitigated, adverse event rates tend to decrease significantly.[9]

---

9 Thomas R. Krause, "Moving to the Second Generation in Behavior-Based Safety," *Professional Safety*, 46 (May 2001), pp. 27–32; *BST Outcome Study* (Ojai, CA: Behavioral Science Technologies, 2002); and John H. Hidley, "Critical Success Factors for Behavior-Based Safety," *Professional Safety* (July 1998): pp. 30–34.

Step 8: Sustain the Leading with Safety process for continual improvement

While leadership activity wields a direct and immediate influence over the safety climate, shifting organizational priorities can readily impede the metamorphosis of climate into culture. Sustaining mechanisms provide a counterweight to ensure needed continuity of intention and action despite priority changes. Such mechanisms provide accountability for keeping safety visible, for identifying and addressing hazards, for ensuring that others are actively engaged in safety, and for maintaining alignment among the senior leaders on the strategic importance of safety for employees and patients alike.

As we have said elsewhere,[10] by mechanism we mean a set of steps or system components that reliably lead to a defined result. Improving safety continuously, rather than in fits and starts, requires establishing ongoing mechanisms that become normal and reliable practice. The eight steps of the Leading with Safety process create such mechanisms. In phase I, leaders generate action plans for continuous safety improvement in their domains. In phase II, step 7, the behavior observation and feedback process supplies an ongoing, continuous improvement mechanism for safety in the working interface.

The process of behavior observation and feedback provides leadership with a new kind of safety data; namely, information about the day-to-day level of exposure to hazards in the working interface. This new information is proactive and points to specific systems and cultural issues that impede safe performance. By responding effectively to this information, leadership reinforces the value of safety in the organization's culture, directly and materially improves organizational safety, and sustains the mechanisms and systems that continuously drive safety improvement.

In addition, a regular and systematic review, led by the senior leadership team, is a necessary sustaining mechanism. The review process looks at the implementation of each step in the Leading with Safety process and ensures its integrity.

10 Thomas R. Krause, *Leading with Safety* (Hoboken, NJ: John Wiley & Sons, 2005), p. 147.

Case history: Exemplar HealthNet

This hypothetical case history of a fictitious healthcare organization is a heuristic device to illustrate the execution of a planned change in safety climate and organizational culture and to demonstrate some of the ways leaders can systematically use the tools presented here to improve both patient and employee safety. This sketch draws on BST's experience with numerous interventions in a variety of settings, including healthcare.

Exemplar HealthNet[11] is a private, not-for-profit, U.S. healthcare system with 10 locations: seven hospitals (275 beds to 550 beds), two large long-term care facilities, and one hospice. National media coverage of medical errors first brought patient safety to the attention of Exemplar's trustees, who embraced the issue for ethical and moral reasons. In addition, a sentinel event occurred in which a patient admitted for a routine surgery received the wrong medication and died. The nurse who administered the medication had overridden the dispensing machine. Investigation of the event revealed that the nurse had been hired because of pressure from a dominant human resources manager, who had overruled the clinical supervisor and insisted on the hire despite evidence of previous professional lapses on the part of the nurse.

Case history—Step 1: Gain alignment

The Exemplar trustees communicated their desire for a strong commitment to patient safety to Exemplar's new chief executive officer, who realized that patient safety should be regarded as a strategic priority and core value. She was concerned about fragmentation and friction between constituencies, poor incident reporting, and Exemplar's less than ideal standing in the community. She also realized that safety generally, and patient safety in particular, afforded the opportunity to build a more unified and positive organizational culture. She hoped to raise Exemplar's visibility and increase its prestige.

Exemplar's safety process was kicked off at a retreat attended by key leaders from among the trustees, corporate executives, and managers of the 10 facilities. At the retreat, leaders learned the

11 "Exemplar HealthNet" is a fictitious name and has no relationship to any of the thousands of Google returns for "HealthNet."

core concepts of organizational safety and formed the intention that safety was to play a strategic role in the organization's future. They reviewed cultural and leadership data, formulated a safety vision, and developed an intervention plan.

Case history–Step 2: Develop a vision

The leaders formulated the following vision statement for safety:

> Measured excellence in patient and employee safety performance will lead our ability to align constituencies and create a positive, cohesive, and cooperative staff of service providers, thereby establishing Exemplar HealthNet as our community's preferred organizational healthcare provider. We recognize that accomplishing this vision requires behavior and systems changes. Specifically, we will see the following changes:
>
> - There will be more vertical and horizontal safety communication, both within treatment teams and among service providers, managers, and administrators.
> - There will be greater cooperation about safety among and between departments, professional groups, and the administration.
> - Safety itself will move from a priority to a value.
> - We will establish behavior reliability within treatment teams, senior leaders, and department heads.
> - Healthcare safety–enabling elements and organizational sustaining systems will be aligned with safety as a strategic priority.
> - Safety climate and organizational culture measures will be taken annually to give senior leaders and trustees an objective measure of the organizational culture and the tone at the top.

Case history—Step 3: Analyze the current state

The Blueprint for Healthcare Safety Excellence (Figure 2–1) identifies the five elements that influence safety performance in an organization:

- Leadership
- Organizational culture
- Healthcare safety–enabling elements
- Organizational sustaining systems
- Working interface

Assessing each of these elements provides a clear and comprehensive picture of the current state of safety in the organization.

Case history—Gather leadership data

As discussed earlier, leaders are embedded in and create organizational culture (see the Safety Leadership Model, Figure 4–1). Leadership characteristics can be measured using the Leadership Diagnostic Instrument (LDI). Figure 9–1 shows LDI data for Exemplar HealthNet's senior leadership team, which included the system's CEO, her direct reports, and the hospital CEOs. The diamonds along the bar indicate the percentile scores of individual leaders. The bar itself is the consolidated score for the group. In both cases the comparison group is several hundred other leadership teams in the database. It is noteworthy that variability across this group of leaders is very high.

Relatively low safety leadership scores are common for leaders in organizations that have not emphasized safety. These data show that the Exemplar leaders' greatest opportunities were across the first six practices. Each leader had his or her own unique LDI pattern of strengths and opportunities.

Interviews with leaders revealed that safety was low on a long list of priorities and was not getting leadership attention. No one, of course, wanted to see adverse events occur, but safety did not occupy a prominent place on the agendas of these leaders. Many had assumed that if everyone just did their job, safety would somehow take care of itself.

FIGURE 9-1. LDI SCORES* FOR EXEMPLAR HEALTHNET (A FICTIONAL HEALTHCARE SYSTEM).

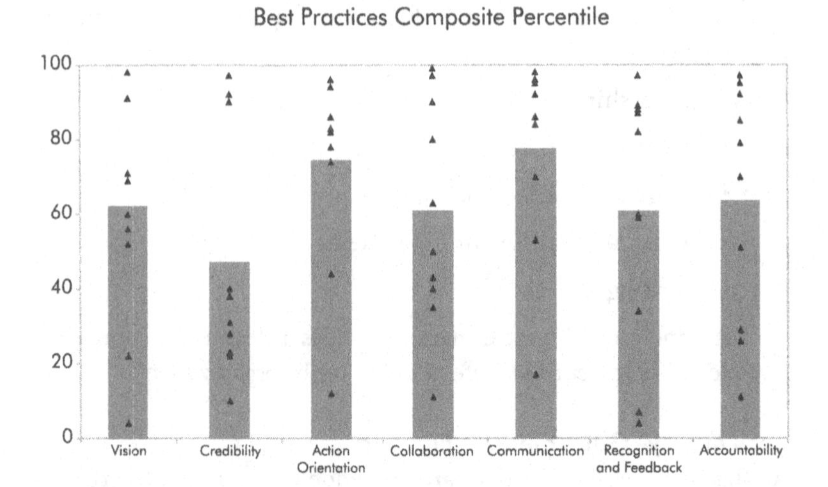

*Triangles indicate individual leadership scores; the bars indicate composite team scores.

Leadership data is best understood in the context of the organizational culture in which the leader works; the culture simultaneously provides the context and is the product of leadership's decisions and behaviors. Cultural data tell the leader where to direct his personal leadership improvement activities.

Case history—Gather cultural data

The Organizational Culture Diagnostic Instrument (OCDI) measures nine dimensions, which predict the frequency of incident events. Exemplar's OCDI data (Figure 9–2) painted a picture of low organizational functioning and average team functioning, a weak safety climate, and significant leadership and supervisory opportunities. In the figure corporate and site data are combined, and the results are given as percentiles[12] in relation to a large database of diverse organizations. These findings were consistent with Exemplar's LDI results.

---

12 Percentile is a rank ordering of the raw score data on a scale of 100 such that the 50th percentile is the middle of the distribution of raw score data.

FIGURE 9-2. CONSOLIDATED OCDI SCORES FOR EXEMPLAR HEALTHNET (A FICTIONAL HEALTHCARE SYSTEM).

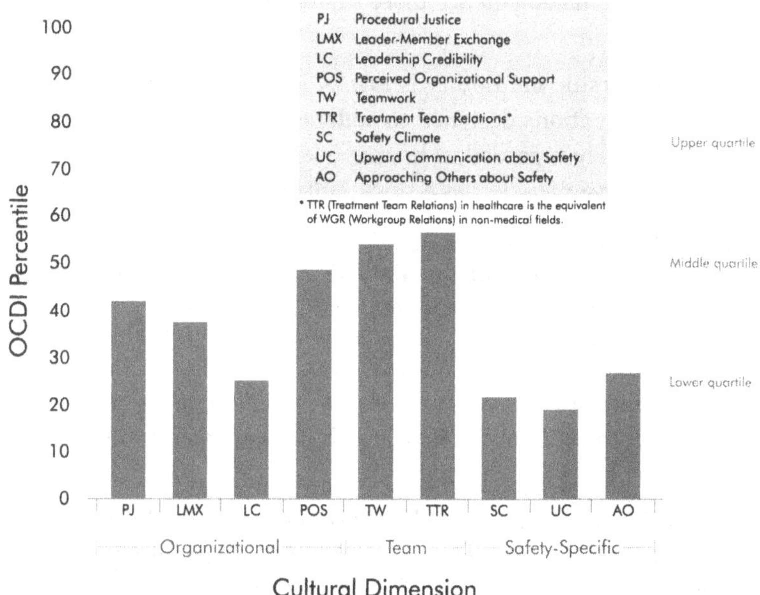

Data from interviews and focus groups confirmed these findings and provided additional information about specific issues that gave rise to Exemplar's difficulties. Here are some highlights from the findings:

- An us-versus-them atmosphere prevails. Physicians express frustration with the nursing staff and with the administration. The administration is seen as bureaucratic and the nursing staff as overburdened and inefficient.
- Nurses express frustration over conflicting demands, shifting priorities, and specific physicians whom they perceive as difficult to work with.
- The administration feels powerless to intervene effectively in treatment-related issues.
- Many leaders, both clinical and administrative, express the view that some number of adverse events are inevitable.

- Systems thinking is rare. Individuals value their ability to perform competently and see breakdowns as the result of others' incompetence more than a systemic issue. Blame is common.
- Leadership credibility is low. A long history of interventions designed to address organizational problems is widely regarded as having been unsuccessful. This negative view leaves a thick cultural residue of cynicism.
- Explicit discussion of safety issues is not part of the organizational culture. Raising a safety issue is often seen as insulting, over-reaching, or futile.
- Overall employee morale is low.
- Teamwork is present in certain departments, where overall functioning is relatively higher than in the organization as a whole.

Organizational culture and OCDI findings varied by Exemplar HealthNet location, level, and department. Enabling systems for both patient and employee safety were partially in place, but these systems were often not well understood by leaders nor actively used by treatment team members. For example, systematic emergency room patient-flow planning had been implemented on paper, but patient flow was actually managed ad hoc, on-the-fly. Employee selection and advancement processes did not give adequate weight to safety performance, and leaders frequently substituted their own decisions for the systematic procedures that were supposed to be used for these functions.

Case history—Gather hazard data

The Exemplar HealthNet interviews, focus groups, OCDI, LDI, and adverse event data all suggested that significant exposure for both employee and patient safety prevailed in the working interface. Employee safety accounted for significant costs.

Case history—Step 4: Develop the intervention plan

The objectives of the safety plan developed by the Exemplar leadership and the measures by which they would evaluate progress included the following:

- Strengthen the organizational culture and safety climate as measured by the OCDI.
- Improve safety leadership skills as measured by the LDI.
- Reduce the frequency of adverse events as measured by a set of three to five metrics to be drawn from existing data.

The leaders' broad strategy began with a set of interventions that would immediately improve the safety climate and create mechanisms for longer-term cultural improvement. The leadership aligned to change the safety climate—i.e., what is "expected, rewarded, and supported at Exemplar HealthNet." The interventions consisted of the following components, whose aim was to accomplish all of phase II by developing the needed skills and installing continuous improvement and sustaining mechanisms:

- Provide leadership coaching for trustees, senior leaders, and selected others based on each leader's personal LDI and 360-degree interview data.
- Establish a clinical leadership role for safety, and engage physicians in safety.
- Provide leadership workshops for less senior organizational and clinical leaders and targeted training based on strategic needs.
- Install a behavior observation and feedback system in the working interface for the benefit of treatment teams and their patients.
- Provide training in systems thinking.

- Develop and align enabling and sustaining systems, especially selection and performance management.
- Monitor the culture by administering the OCDI annually.

The distribution of these interventions across the organization's levels is shown in Table 9–1.

Intervention: Executive and leadership assessment and coaching

Each Exemplar HealthNet senior leader and selected other leaders received a safety leadership assessment using a 360-degree diagnostic survey (the LDI) and, in some cases, interviews of people surrounding the leader. The interviews revealed that leaders were perceived as practicing a hands-off style of leadership until a problem arose, at which time they were seen as quick to assign blame. Focus groups revealed that leaders were seen as especially unwilling to take action on specific known systems issues and unwilling to deal with certain problematic physicians.

In their coaching sessions (see the leadership coaching sidebar), the leaders developed individual action plans to improve their leadership capabilities and support the overall initiative. After reviewing their LDI and OCDI results they were eager to adopt new best practices in their leadership style and leadership behaviors to optimize their impact on their organization's culture. To accomplish this change in behavior, they sought to better understand the relationships among best practices, leadership style, and OCDI dimensions and to identify relevant behaviors with which to enact best practices.

Intervention—Physician engagement

The phase I gap analysis at Exemplar HealthNet revealed that physician engagement in patient safety was at best passive and reactive. Since successful comprehensive culture change in healthcare requires physician leadership and participation, physician engagement became a formal part of the intervention plan. Physicians—whether they know it or not, and regardless of whether they want the role—are organizational leaders and culture shapers in the healthcare delivery industry. The culture of healthcare is largely authority-based, and physicians remain the leading authorities.

Enlisting Exemplar physicians as leaders consisted of helping them understand what they were already doing to create culture and helping them find ways to shape culture with specific objectives in mind. Physicians invariably favor improved patient safety but often believe they have neither the expertise nor the time to lead an effort they see as not their responsibility. Once they understand how they are already creating culture, it is easier for them to assume responsibility for doing so deliberately.

TABLE 9-1. INTERVENTIONS AND RESPONSIBILITIES, BY GROUP, FOR EXEMPLAR HEALTHNET (A FICTIONAL HEALTHCARE SYSTEM).

	TRUSTEES AND SENIOR CLINICAL/ ADMINISTRATIVE LEADERS	PHYSICIANS, DEPARTMENT HEADS, MANAGERS, AND SUPERVISORS	OTHER TREATMENT TEAM MEMBERS AND EMPLOYEES
Interventions	▪ Executive coaching ▪ Dialogue skills training ▪ Training in transformational leadership skills	▪ Leading with Safety workshops ▪ Training in transformational leadership skills ▪ Leadership coaching (selected leaders) ▪ Training in systems thinking	▪ Leading with Safety workshops ▪ Behavior observation and feedback ▪ Dialogue skills training
Responsibilities	▪ Oversee systems alignment and improvements	▪ Implement systems alignment and improvements	▪ Implement systems alignment and improvements

This part of the Exemplar HealthNet intervention included the appointment of physician mentors to help peers adopt best practices, to develop and share ideas for practice quality improvement, and to provide positive coaching for poorer performing physicians (those who were disruptive or abusive or under disciplinary review).[13]

13 See, for example, the University of California, San Diego School of Medicine's Physician Enhancement Program (PEP), part of its larger Physician Assessment and Clinical Education (PACE) Program, at www.paceprogram.ucsd.edu.

SIDEBAR. LEADERSHIP COACHING.

FROM REMEDIAL TO DEVELOPMENT

Not that long ago, executive coaching was viewed as a remedial strategy; when executive coaches were engaged, it was usually to support executives who were struggling with leadership or relationship issues. Over the past decade or so, however, a remarkable—and positive—transformation has taken place. What was once seen as a means of "fixing" broken executives (the grown-up equivalent of being sent to the principal's office) has become a highly valued executive perk. Coaching has become an important tool for leadership development. Many organizations now assign a coach to all senior leaders and, in some cases, many of their high-potential, upper-level middle managers. Healthcare organizations are beginning to avail themselves of this tool as well.

With its transition to a proactive development strategy, leadership coaching has had to adapt to meet the needs of leaders seeking to enhance their overall effectiveness. With remedial coaching, the issues are generally apparent, and the coach is usually engaged to help the leader address the specific behavior or behaviors causing difficulties. With developmental coaching, the coach becomes the voice of the organization, helping the leader understand how his portfolio of behaviors sways key stakeholders and either advances or impedes the organization's ability to meet its strategic goals. While this may include modifying less constructive behaviors, it also frequently includes helping the leader leverage his strengths to greater advantage.

LEADING FROM THE BEHAVIOR LEVEL

One of the reasons leadership often seems mysterious is that it is frequently discussed at the characteristics level; leaders are characterized as "charismatic," "compelling," "visionary," or even "Machiavellian." Although describing leaders in terms of their alleged characteristics may be interesting, it sheds little light on how or what leaders can do to improve as leaders. Telling someone she needs to be more charismatic or visionary does little to clarify which behaviors she needs to change in order to improve her day-to-day leadership.

A characterization is a perception or generalization that people develop through observation of a leader's behaviors. If we address the behaviors themselves, rather than the characterization, we can begin to help leaders think about how to change their behaviors to enhance their overall effectiveness.

An experience with a senior leader illuminates this point. In confidential interviews on behalf of the leader, his direct reports frequently characterized him as "indecisive." As we probed further, we learned that employees, when attending meetings in which decisions were anticipated, observed that the executive would listen quietly and with apparently minimal engagement. When the meeting time expired, he would simply thank everyone and end the meeting. The attendees, having anticipated that a decision would be forthcoming in the meeting, interpreted the leader's behaviors as indecisive.

Just informing the leader that he was perceived as indecisive would not have been constructive. In fact, it could have been counterproductive if he responded by making ill-informed decisions just to address the perceptions of others. But once the behaviors that led his reports to characterize him as indecisive were identified, the leader was able to develop a simple plan to change the way others viewed him.

Henceforth, he opened meetings by identifying and articulating the decision to be made, how it would be made (i.e., by consensus, by majority vote, by the leader with input, or by the leader without input), and what information was required to make the decision. Most important, he articulated clearly when the decision would be made. In other words, rather than trying to become more decisive, he became more definitive in communicating his decision-making process.

The effect was almost immediate: Follow-up interviews with his direct reports a month later indicated that, by simply becoming more transparent in decision making, he was able to change the perception of himself from "indecisive" to "very decisive."

CAN PEOPLE REALLY CHANGE?

Can working with a coach help a leader change behaviors that have been developed and ingrained over years? The answer is yes, of course. People change their behaviors all the time. If you drive to work the same way for 30 years and a faster route becomes available, it is easy to change your 30-year routine. It is more difficult to change the underlying value you hold that led you to evaluate your behavior and decide to change it. You did not change the value of wanting to get to work as quickly as possible; you changed your behavior because you found a new behavior that gave you an outcome that better supported your value. The coaching process relies on this dynamic to help leaders quickly accomplish the behavior change to which they aspire.

Intervention—Leading with Safety workshops and targeted training

Modeled after the Patient Safety Academy, these one-day events were designed to cover the basic concepts, research basis, and leading-edge thinking that goes into patient and employee safety. The agenda covered the following topics:

- Why is safety an ethical imperative, a core value, and a strategic priority at Exemplar HealthNet?
- What controls safety in our organization? (The Blueprint for Healthcare Safety Excellence)
- What characterizes a culture that values safety? (The OCDI)
- What characterizes outstanding safety leadership? (The LDI)
- Why does safety require change at the individual behavior level, and how can we accomplish that? (Our intervention plan)

The gap analysis identified specific training needs. A variety of training was provided to various Exemplar groups on topics such as cultivating dialogue skills, strengthening transformational leadership skills, and understanding systems thinking.

Intervention—Training in systems thinking

Systems thinking—the ability to see how processes, people, and technology interact to produce outcomes—was essential to the improvement in patient safety that Exemplar desired. In workshops, physicians, department heads, and select managers and supervisors were helped to understand that their responsibility was to participate in systems that ensure safe outcomes for patients and employees. Training in root cause analysis, ongoing participation in routine investigative analyses after each adverse event, and behavior observation and feedback were among the methods that created this understanding.

Intervention—Behavior observation and feedback

Establishing an ongoing behavior observation and feedback system at Exemplar HealthNet supplied the intervention plan's needed safety-sustaining mechanisms and plenty of meaningful engagement opportunities. Creating a feedback-rich environment in the working interface accelerated culture change. Engagement opportunities for treatment team members included involving them in identifying safety-critical behaviors and inviting them to participate as peer observers who reinforced fellow team members' safe performance of critical behaviors, thereby providing new consequences for safety. This process generated discussions about safety where none previously existed. Cumulatively, involvement in these activities hastened the culture's assimilation of safety-critical behaviors, which rapidly became "how we do things around here."

Intervention—Alignment of enabling and sustaining systems

The Exemplar gap analysis found that many of the needed enabling systems were present but not well used, and that many sustaining systems were lacking. The intervention strategy vigorously built leadership and culture while deploying system improvements in support of culture building.

Intervention—Periodic cultural reevaluation

The Exemplar HealthNet safety plan calls for readministering the OCDI in six months and then, depending on the results, annually thereafter. This aspect of the plan will provide a measure of progress, supply an important complement for a safety early warning system, and give leaders and trustees the information they need to monitor and manage the tone at the top.

When the Exemplar HealthNet leaders completed the Patient Safety Academy, they had a set of plans for moving forward, both as individuals and as a team, to create a strong organizational culture and safety climate. They had a high-level plan for achieving the leadership team's safety vision. This plan was then fleshed out, implemented, measured, and adjusted. The new data it generated were used to guide adjustment and replanning. This action cycle constituted a continuous improvement safety process for Exemplar.

In this chapter and the previous one we outlined an eight-step process for leading effective changes in organizational culture to improve safety and illustrate some of the ways leaders can use the tools explained in this book. In the next chapter we will explore how this process helped NASA change its climate and culture after the space shuttle *Columbia* tragedy in 2003.

CHAPTER TEN

NASA AFTER COLUMBIA: *LESSONS FOR HEALTHCARE*

NASA AFTER *COLUMBIA:* LESSONS FOR HEALTHCARE

Following the tragic loss of the space shuttle *Columbia* and its crew of seven on February 1, 2003, the National Aeronautics and Space Administration (NASA) undertook a major initiative to transform its organizational culture and safety climate. BST was selected to design and support the initiative. The authors and a team of colleagues worked intensively on the project over an approximately two-year period ending in June 2005. Near the end of this engagement, Sean O'Keefe,[1] who served as NASA Administrator during the *Columbia* tragedy and its aftermath, published an op-ed piece[2] summarizing what had been accomplished:

USA TODAY
Monday, April 25, 2005
By Sean O'Keefe

Step by step, NASA is doing what it takes to "fix the culture"

One small step for NASA could turn into a giant leap for the space program.

In 2003, the Columbia Accident Investigation Board found that NASA's safety culture contributed as much to the space shuttle *Columbia* accident as any mechanical failure. The board assessed that we needed to "fix the culture"—and thus was born one of the largest, most complex organizational changes ever.

---

1 Sean O'Keefe, NASA's 10th Administrator, served from 2001 to 2005.

2 Reprinted by permission of the author.

During the past 13 months, the agency has developed a model for understanding culture and how it changes. It has measured specific aspects of the culture across all 19,000 employees, designed and tested an intervention method, and is implementing the strategy that will reach through NASA by the end of 2005.

While activity cannot be equated with achievement, the fact is, the program is already working. A recent survey taken in February of 2004 and then October [2004] found "solid, measurable progress" from the initiative. It looked at nine fundamental attributes of culture—including employee-supervisor relationships, management credibility, teamwork, work-group relations, the safety climate, and upward communications—and each category showed improvement. Employees are increasingly more comfortable raising safety questions, and their concerns are being fully explored.

"Measurable progress"

Statistically and anecdotally, the progress is real—and the process is working faster than what has been achievable in many organizations. NASA set a very aggressive schedule, and the agency's mind-set for achievement has helped to create solid, measurable progress.

This does not mean that NASA's culture is fixed. Organizations don't fix cultures the way plumbers fix leaks; they address a set of issues so that a new culture develops. In the tough world of organizational change, that's progress, not success. At NASA, some people are embracing change, others are awaiting their turn and still others will never buy in—that's human nature.

Of course, there are those within the agency and in the space-exploration community who still aren't convinced. *USA Today* recently reported the views of those who think NASA's culture has not been fixed. Debates have emerged, as reported Friday in the *New York Times*, on the methods for measuring the actions taken to meet the recommendations of the Columbia Accident Investigation Board. Such opinions show the dialogue is continuing.

The relationship between NASA employees and their work is unusual. NASA ranks high in employees' ratings of workplace desirability. NASA employees are not as motivated as others by financial gain or other extrinsic rewards. They are motivated by their personal connection to what the agency does, what it wants to do, and what it means to be part of space exploration and the discovery of the universe.

Not just a job

Space exploration is different from other pursuits. It has an inherent worth, giving meaning to the tasks the organization performs. It creates passion for the goals the organization is trying to accomplish. These factors extend to the general community interested in space exploration, and to every classroom in America. NASA's success means something to us as Americans. In turn, everyone has an opinion about what should happen at NASA and whether it's happening fast enough.

The fact is, NASA exists to take on bold adventures for the sake of education, science and exploration, national security and native curiosity. The initiative to change the culture is designed to help NASA manage the risks and deal with the challenges. When the shuttle returns to flight next month, or whenever, the public should be confident that the process is managed by diligent professionals cognizant of the risks they are charged to manage.

The ongoing progress of NASA is worthy of recognition. Scrutinizing NASA may be a national pastime, but a safer, successful space program should be considered a national treasure.

In this chapter we use BST's work with NASA to illustrate how leaders can employ the tools and the change process presented earlier in this book to build a strong safety climate and lead intentional change in organizational culture. The healthcare community understands that it must address patient safety *organizationally* and that sustained safety improvement requires "a culture of safety."[3] Strong parallels exist between the safety challenges healthcare organizations face today and those faced by NASA. As you read this case history, notice the similarities with

3 *To Err Is Human: Building a Safer Health System*, eds. Linda T. Kohn, Janet M. Corrigan, and Molla S. Donaldson (Washington, DC: National Academy Press, 1999). Viewed at www.nap.edu. *See also* H. S. Ruchlin, N. L. Dubbs, M. A. Callahan, and M. J. Fosina, "The Role of Leadership in Instilling a Culture of Safety: Lessons from the Literature," *Journal of Healthcare Management*, 49 (January–February 2004): pp. 47–59.

your own organization as well as the systematic way the culture change process is applied. We will discuss the implications for healthcare after the case study.

NASA's approach to culture and climate transformation

As Sean O'Keefe wrote in his op-ed piece, the Columbia Accident Investigation Board found that NASA's culture and related history contributed as much to the *Columbia* accident as any technical failure. After a six-month investigation, the board reported the following:

> The organizational causes of this accident are rooted in the Space Shuttle Program's history and culture, including the original compromises that were required to gain approval for the shuttle program, subsequent years of resource constraints, fluctuating priorities, schedule pressures . . . Cultural traits and organizational practices detrimental to safety were allowed to develop, including: reliance on past success as a substitute for sound engineering practices... organizational barriers that prevented effective communication of critical safety information and stifled professional differences of opinion; lack of integrated management across program elements; and the evolution of an informal chain of command and decision-making processes that operated outside the organization's rules.[4]

Although the board made no recommendations explicitly addressing leadership practices, its report identified many examples of gaps in the leadership practices that support safety, such as:

- Failing to follow NASA's own procedures
- Requiring people to prove the existence of a problem rather than assuming the need to ensure there was not a problem
- Creating a perception that schedule pressure was a critical driver of the program

---

4 Columbia Accident Investigation Board, *Columbia Accident Investigation Board Report* (August 2003), vol. 1, p. 9.

As a result of the board investigation and related activities, NASA established the objective of completely transforming its organizational culture and safety climate. At a minimum, it targeted making measurable progress in changing its culture within six months and having broad changes in effect across the agency in less than three years. NASA identified the six-month marker as particularly critical as the agency prepared to return to flight.

After reviewing proposals from more than 40 organizations, NASA selected our firm to assist in the development and implementation of a plan for changing the culture and the safety climate. We were asked to design and initiate a systematic, integrated, NASA-wide approach to understanding the prior and current norms of the safety climate and culture, and to diagnose aspects of climate and culture that would not support the agency's effective adoption of changes identified by the Columbia Accident Investigation Board. We were further asked to propose a course or courses of action to change behaviors and to introduce new norms that would:

1. Eliminate barriers to a "safety culture"[5] and mindset
2. Facilitate collaboration, integration, and alignment of the NASA workforce in support of a strong safety and mission success culture
3. Align our work with current initiatives already under way in the agency

NASA asked us to assess their current status and develop an implementation plan within 30 days. Initial results at three NASA locations were measured in September 2004 and interventions put in place immediately thereafter. Reassessment was completed in February 2005. Based on the success achieved at those locations, the remaining eight NASA locations received the same set of assessment and intervention.

---

5 As noted earlier (see chapter 2) "safety culture" is a bit of a misnomer. *Safety climate* is one of the nine dimensions of *organizational culture*.

Assessing NASA's existing culture and climate

We approached the task of assessing NASA's culture with the belief that much of it was positive. Our challenge was to identify and build on these positive aspects, strengthen the overall culture, and at the same time address the issues raised in the board report. In the aftermath of the *Columbia* tragedy, a strong safety climate was already developing (as often happens in organizations after a sentinel event). However, we were concerned that in the absence of properly focused efforts, the climate change would not be sustained, the culture would not change, and over time the new safety climate was likely to be compromised by the inevitable scheduling, budgetary, and operational pressures that occur in any organization.

Organizational Culture Diagnostic Instrument (OCDI) findings

We administered a specially modified version of our OCDI at all 11 NASA locations via a Web-based link to solicit information about mission safety. In addition to the basic survey scales, we added questions specifically designed for use in NASA to evaluate the current situation in comparison to the desired state and to gather data on several specific culture-related issues raised by the board report. The overall survey response rate was 45.2%, and our tests indicated that the responding group was comparable to the overall NASA population. Agency-wide response to the basic survey scales is shown in Figure 10–1 (percentile scores) and Figure 10–2 (raw scores). Figure 10–1 compares NASA with the normed database that BST has compiled with this survey.

On the agency-wide level, NASA scored well in relation to other organizations in our database on most of the scales making up the survey. NASA scored above the 90th percentile on approaching others and workgroup relations, and between the 80th and 90th percentiles for teamwork and leader-member exchange. These results indicated that across the agency there was generally effective team functioning at the local level, with employees who have the ability and inclination to speak up to peers.

NASA scored lowest on two scales: perceived organizational support (46th percentile) and upward communication (62nd percentile). As you will recall from chapter 3, perceived organizational support measures employees' perceptions about the organization's concern for their needs and interests. These perceptions in turn influence beliefs about the organization's values for safety. This result influences employees' willingness—or unwillingness—to raise safety concerns. Upward communication measures perceptions about the quality and quantity of upward communication about safety, the extent to which people feel encouraged to bring up safety concerns, and the level of comfort discussing safety-related issues with the supervisor.

Lower scores on perceived organizational support and upward communication indicated areas for particular attention during the culture change effort. Senior management and the behaviors they stimulate through the management chain influence both of these dimensions, which also have a strong influence on the culture in ways that relate directly to mission safety.

FIGURE 10-1. COMBINED OCDI PERCENTILE SCORES FOR ALL NASA LOCATIONS PRIOR TO THE INTERVENTION.

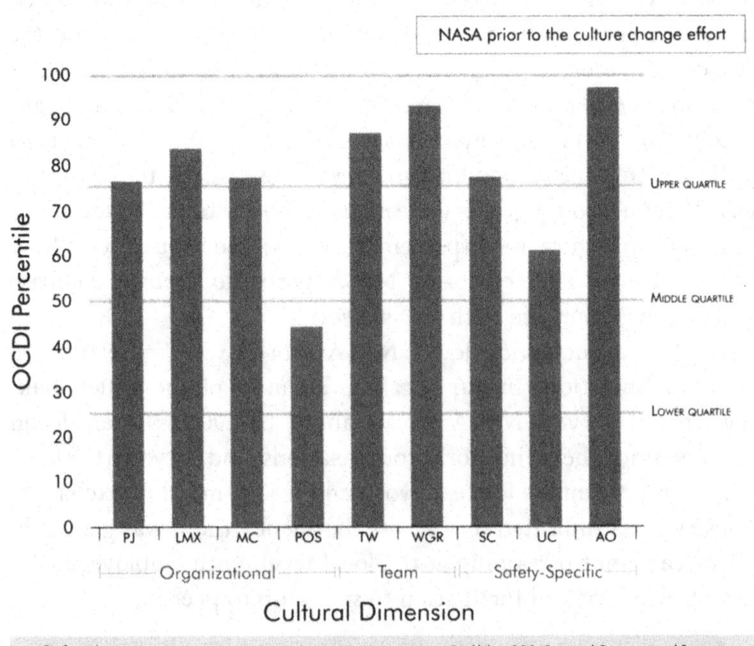

PJ · Procedural Justice LMX · Leader-Member Exchange MC · Management Credibility POS · Perceived Organizational Support
TW · Teamwork WGR · Workgroup Relations*
SC · Safety Climate UC · Upward Communication about Safety AO · Approaching Others about Safety

* WGR (Workgroup Relations) in non-medical fields is the equivalent of TTR (Treatment Team Relations) in healthcare.

FIGURE 10-2. COMBINED OCDI RAW SCORES FOR ALL NASA LOCATIONS PRIOR TO THE INTERVENTION, WITH MEAN AND STANDARD DEVIATION.

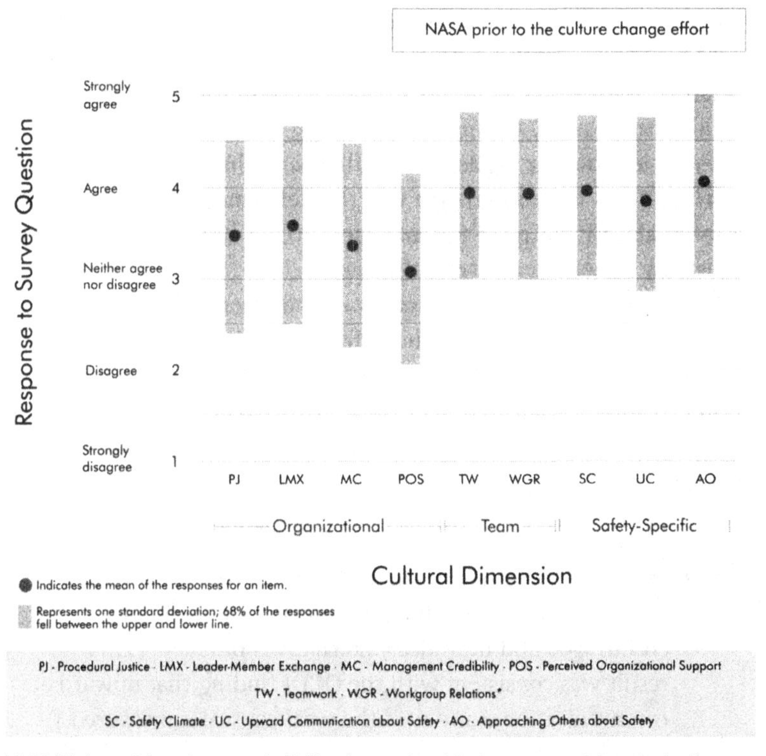

* WGR (Workgroup Relations) in non-medical fields is the equivalent of TTR (Treatment Team Relations) in healthcare.

Interview findings

To help supply context and depth for the survey results, we conducted a series of interviews and focus groups with 120 people at representative locations. At each location we interviewed individual members of senior management and met with representative groups of individual contributors and groups of supervisors and managers. The interviews disclosed a strong sense of dedication and commitment to the agency's work. However, we also found frustration about obstacles to upward communication and deficits in leadership skills.

We heard in several ways that there were impediments to speaking up at NASA. On more than one occasion individuals hung back at the end of our group sessions and made comments after others had left or wrote notes to us expressing thoughts they had not brought up in front of others. These comments usually described barriers to communication. This feedback was consistent with the upward communication survey result and indicated that some non-managers within NASA felt that open communication was impeded. We also heard many comments indicating that not all managers and supervisors had the leadership skill levels that many considered appropriate. A common theme was the deficiency of respect for the individual and the need for managers to act more consistently in ways that better demonstrate this value.

This assessment information was consistent with the themes NASA had obtained during the Safety and Mission Success Week that was held two months prior to our engagement in which each center director collected feedback from his workforce on the Columbia Accident Investigation Board report. The issues raised included:

- Workers lacked a process for delivering upward feedback.
- Minority opinions needed to be embraced to create an open atmosphere in which disagreements would be encouraged and new ideas/alternatives pursued. (This result was consistent with the OCDI finding that upward communication was one of the weakest scales measured.)
- Leaders did not follow words with actions. (This tendency contributed directly to lower scores on management credibility.)
- Leaders also delivered a message of "what" without delivering the "why." (This gap likely contributed to NASA's lower scores for management credibility and perceived organizational support.)
- The workforce wanted and needed a culture that values and promotes respect and cooperation. (This need related to perceived organizational support.)
- Contractors were treated as second-class citizens. (This treatment resulted in inhibited communications, which has the potential to impede performance excellence.)

Assessment conclusions

NASA defines its values as follows: People, Safety, The NASA Family, Excellence, and Integrity. Our assessment found that the NASA culture had a long legacy of taking a can-do approach to task achievement, but did not yet fully demonstrate the agency's espoused values of Safety, The NASA Family, Excellence, and Integrity. The culture revealed an organization in transition, with many ongoing initiatives and the lack of a clear sense at working levels of how it all fits together. Examining NASA's espoused values, we found that:

- *Safety* was something to which NASA personnel were strongly committed in concept, but NASA had not yet created a culture that was fully supportive of safety. Open communication was not yet the norm, and people did not feel fully comfortable raising safety concerns with management.
- *The NASA Family* value, though espoused, was inconsistent with the fact that people felt disrespected and unappreciated by the organization. As a result, the strong commitment people felt to their technical work did not transfer to a strong commitment to the organization. NASA personnel in support functions frequently did not fully understand or appreciate their connection to the agency's mission, and people in technical positions did not fully value the contribution of support functions to their success.
- *Excellence* was a treasured value when it came to technical work, but it was not seen by many NASA personnel as an imperative for other aspects of the organization's functioning (such as management skills, supporting administrative functions, and creating an environment that encourages excellence in communications).
- *Integrity* was generally understood and manifested in people's work. However, there appeared to be pockets in the organization in which the management chain had sent

> signals—possibly unintentionally—that raising negative issues was unwelcome. This was inconsistent with an organization that truly values integrity.

In summary, our assessment identified a clear opportunity to help NASA become an organization that lives its values.

BST's NASA intervention

Based on our assessment, we recommended that the culture change initiative build on the strengths identified in the safety climate and culture survey. NASA employees generally worked well as teams, liked and respected each other, and felt comfortable talking to peers. These strengths could be harnessed to create reinforcement mechanisms for behaviors that support the agency's values and desired culture.

We recommended that the culture change initiative concentrate on helping managers and supervisors maintain an effective balance between task orientation and relationship orientation. At NASA, many managers had a natural task orientation, which is not unusual for technical organizations. However, strong task orientation at the expense of relationships can lead to the inhibition of upward communication and to weakness in perceived organizational support. By taking steps to help managers and supervisors improve their balance between task orientation and relationship orientation, NASA could move toward integrating its values of Safety and People and create a culture that would support the agency's mission more effectively.

We knew the NASA culture tended to think in terms of identifying problems and solving them through discrete projects. We believed NASA needed to avoid falling into the organizational trap of viewing its response to the board report purely in a project-driven manner. If NASA were to create separate projects to address specific issues in the report, the agency could fail to address the underlying cultural issues that gave rise to many of the problems in the first place. This inclination perhaps explained why safety climate changes observed after previous accidents (e.g., the space shuttle *Challenger* accident) had failed to take hold and become part of the ongoing culture.

To address NASA's needs and build on its strengths, we developed a culture change plan based on one core concept: *Organizational values must underlie the definition of desired culture.* A favorable safety climate in the wake of a serious incident provides a limited window in which culture change is easier than it might otherwise be. However, such a safety climate usually departs significantly from the culture that produced the incident. A new climate inconsistent with the long prevailing culture is not sustainable without effort. Mechanisms are needed to sustain the climate change and to shift the culture.[6]

We and NASA concluded there should be one, single culture change initiative that incorporated all its culture-related projects. NASA was in a period of change, with many active teams and task forces. Many of these task forces had identified cultural issues, and the activity of so many task forces raised the possibility that there could be overlapping or contradictory initiatives.

We designed a six-month plan to begin the culture change at three locations while validating the approach to fit NASA. The three locations—the Glenn Research Center, the Stennis Space Center, and two large directorates of the Johnson Space Center (Engineering and Mission Operations)—collectively comprised approximately 3,600 people.

Vision of NASA's future state

At the outset, NASA's senior leadership reexamined the organization's core values and reaffirmed those to which the agency aspired. Those values helped NASA articulate a vision of the future state that would exist following successful culture change:

> The objective of this effort is to strengthen the organizational culture and safety climate at NASA. In this desired future state, each individual feels highly valued as an individual and knows that his or her contributions are appreciated. Everyone at the agency, in all roles and at all levels, understands the important ways they contribute to the agency's exciting mission, feels like an integral part of the larger agency team, understands the way that others

6 See the discussion of safety-enabling mechanisms in chapter 2.

> contribute to the larger team effort, and is committed to the success of the agency and its overall mission. Managers and executives at every level of the agency, from top to bottom, routinely treat people with respect. People are comfortable in raising issues, and confident that the issues raised are considered and appropriately factored into decisions. There is a high level of trust in management, and a sense that management, in turn, trusts each individual.
>
> In this desired future state, safety is widely recognized as an integral component of mission success and is considered by every individual in everything they do. The agency is recognized for its pursuit and outstanding achievement of cutting-edge endeavors as well as its extraordinary safety record, all of which are understood as compatible goals.

In designing a strategy to achieve the culture change objective, we began with the recognition that culture expresses shared perceptions, beliefs, and behaviors. It embodies unstated assumptions. We knew that if we changed those perceptions and beliefs, we would change the culture.

Individual perceptions and beliefs are influenced by a variety of factors subject to intervention. For example, perceptions and beliefs about the organization are strongly influenced by individuals' interactions with their immediate supervisors. A change in the leadership behavior of the immediate supervisor influences culture, but this change is unlikely to occur with sufficient consistency and regularity unless there are changes in the leadership behavior of that supervisor's supervisor. Therefore, we needed to change behavior upward through the leadership chain of command—through division chiefs, directors, and center directors.

The new conduct we wanted to encourage in NASA's first-line supervisors—branch chiefs—included a set of critical behaviors that exemplified NASA's core values, behaviors needed to close the gap between the current state and the leadership safety vision, such as communication, consideration for individuals, management consistency, and decision making (Figure 10–3). Together with NASA we reviewed the behaviors at each location where the culture change effort was to be implemented. This review verified the relevance of the behaviors to each location and developed examples of how each behavior would be manifested at the location, to embellish the definition for local use (Figure 10–4).

We designed a multipronged approach of specific activities that included leadership coaching for senior leaders, a behavior observation and feedback process, and multirater feedback and leadership skills training for all leaders (Figure 10–5). A communications campaign was also launched at each location to inform people about the changes occurring.

FIGURE 10–3. FIRST-LINE SUPERVISOR CRITICAL LEADERSHIP BEHAVIORS AND THEIR RELATIONSHIP TO NASA'S LEADERSHIP SAFETY VISION.

FIGURE 10-4. IMPLEMENTATION STRATEGY FOR INDIVIDUAL NASA LOCATIONS.

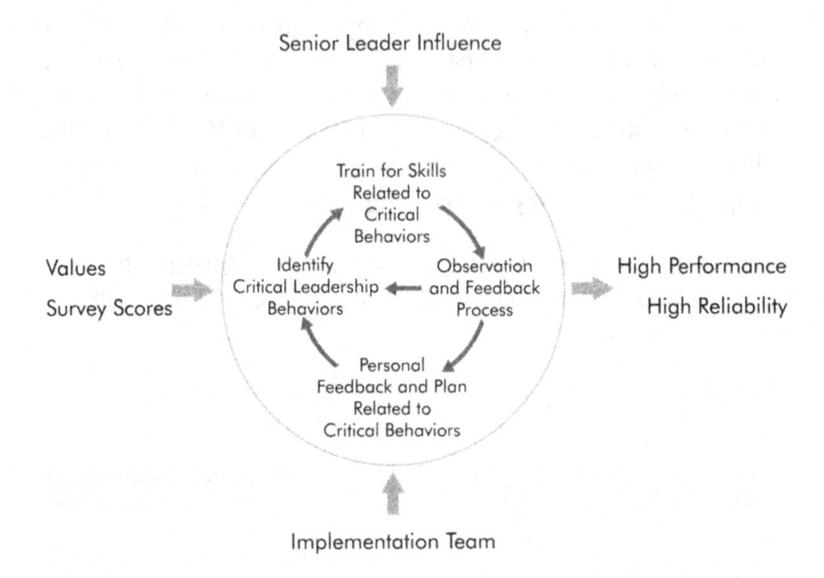

FIGURE 10-5. COMPONENTS OF THE CULTURE CHANGE PROCESS AT NASA CENTERS.

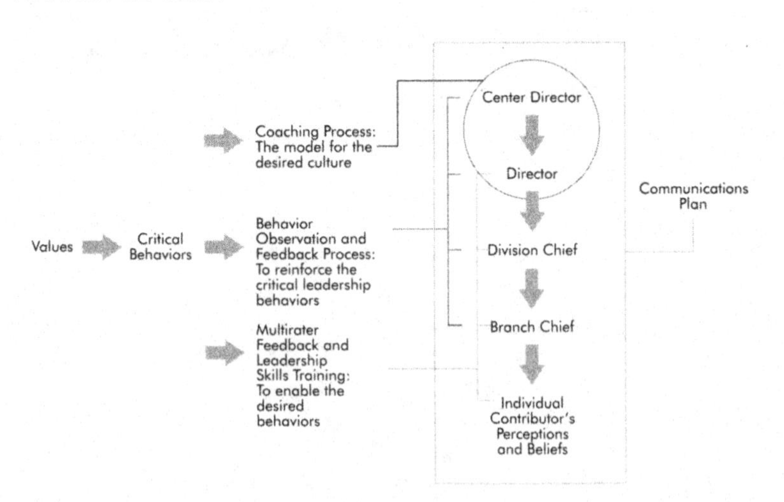

Leadership coaching

To help NASA's senior-most leaders support the culture change, we employed a leadership coaching process that helped them improve their ability to support the critical behaviors plus practice these behaviors themselves (also see the leadership coaching discussion in chapter 9). These organizational leaders had an important but indirect influence on the perceptions and beliefs of most individual contributors and needed to possess strong leadership skills and a solid understanding of how they exert influence. It was important that they set the direction for the desired culture through everything they did and that they learn to create consequences that caused their reports to do the same. Accordingly, the coaching process was designed to help senior leaders understand their leadership strengths and weaknesses.

The process began with a detailed individual assessment, including a 360-degree diagnostic survey (see the discussion of the Leadership Diagnostic Instrument, or LDI, in chapter 4) and interviews with subordinates, peers, and managers. The assessment resulted in a detailed feedback report that gauged the individual's leadership style and best practices. Because this report employed information from coworkers familiar with each leader and supplied detailed examples of his or her leadership behavior, it filled a vacuum that most senior leaders experience—*a lack of direct feedback on their leadership.*

The coach then reviewed the feedback report with the leader and helped him or her develop a coaching action plan. This plan identified areas for the leader to concentrate on, drawing on the critical behaviors, the actions needed to drive support for NASA's values, and the leadership best practices. Once the plan was developed, the coach gave the leader guidance in implementing it.

The coaching process was used with senior leaders, beginning at the top of the agency and extending down through the management chain to the senior-most levels of each center. Leaders eager to acquire or strengthen specific best practices especially appreciated the coaching, but it was also well received by those who were less motivated. None of the leaders reported that it was intrusive.

Behavior observation and feedback

All leaders in the organization were asked to adopt and consistently use the critical leadership behaviors. A process of behavior observation and feedback promoted these behaviors. Leaders received regular, structured, positive reinforcing feedback on their use of critical behaviors and guidance feedback on missed opportunities to use these behaviors. Their use of critical behaviors was encouraged by those senior to them in the organization.

Observers were NASA employees, trained by BST. Anonymous data were gathered during these observations, allowing the local implementation team to track progress in promoting critical behaviors and to analyze the reasons for nonperformance. These analyses were then used in removing barriers to using critical behaviors.

Multirater feedback

To help leaders understand which types of behavior represented strengths, and which represented areas for improvement, we gave each leader his or her LDI scores. (Recall from chapter 4 that the LDI provides feedback from peers, subordinates, and superiors on the leader's degree of transformational leadership and use of leadership best practices.) Leaders attended a workshop to review and discuss the results and to develop individual action plans aimed at increasing their use of leadership behaviors that supported the organization's values.

Skills training

The objective of the skills training was to improve the leaders' ability to perform the critical behaviors and support the desired culture. Managers received two days of training, which covered cognitive bias awareness and feedback skills (on day 1) and leadership skills such as building trust, valuing minority opinion, and influencing others (on day 2). Each training segment explicitly addressed the critical behaviors being promoted in the culture-change initiative.

Communication

Communication was the fifth element of the process for altering the culture, a challenge that had two aspects: *what* was occurring and *why* it was occurring, both of which had to be communicated at the outset. Then, as the implementation proceeded, it was especially important to communicate early progress. The specific mechanisms for the communication varied from center to center based on the communication vehicles available locally. Site newsletters, intranets, and all-hands meetings helped relay information. In addition, managers were encouraged to talk about the implementation progress at their staff meetings.

Results at NASA

For five months beginning in mid-April 2004, we worked with the Glenn Research Center, the Stennis Space Center, and the Engineering and Mission Operations Directorates of the Johnson Space Center. This initial phase of work was designed to learn how best to deploy the culture change approach while meeting the objective of achieving measurable progress in six months.

Soon after implementation began, various results became apparent. We started hearing anecdotal evidence that gave early indications that the culture change effort was beginning to work:

- A division chief reported: "I wasn't sure of this thing in the beginning. Now I am convinced that it will help us; we need to support it. I have invited observers to my meetings; I encourage you to do the same."
- An observer reported: "Helps me be less judgmental and see myself as others do."
- A division chief asked that two of his meetings be observed.
- A branch chief reported: "I found myself conducting my branch meetings and day-to-day interactions differently as part of this effort. I am convinced that others will also change their habits; even if they are not bad right now, but improvement is good."

- One implementation team had a well-known skeptic as a member. After observer training, he got up and told the group that he hadn't been in favor of this, but now that he understood it, he thought it was going to make a big difference.
- Individuals asked for 360-degree leadership surveys to give them feedback.
- Training evaluations consistently indicated that participants arrived as skeptics and left as believers (they switched from "prisoners" to "advocates").
- Division chiefs gave each other feedback in a staff meeting, referring to the coached behaviors.
- An observer was invited to attend a mission management team meeting. An invitation to this critically important forum signaled that senior leaders valued the observation and feedback process.

As data began to accumulate from the behavior observation and feedback process, we started seeing improvement in the frequency with which a critical behavior was performed rather than the opportunity missed. The data produced by the process (not shown) became a mechanism for identifying where next to place the emphasis to achieve further improvement.

About six months after the start of the culture change effort we administered the specially modified OCDI again. We used the same instrument as in the initial assessment effort as well as the same methodology. All scales on the basic Safety Climate and Culture survey showed significant improvement over the pre-implementation scores. Results for the Glenn Research Center are shown in Figures 10–6 and 10–7, for the Stennis Space Center in Figures 10–8 and 10–9, and for the Johnson Space Center in Figures 10–10 and 10–11.

In addition to the basic scales, this survey once again included a series of NASA-specific questions grouped into thematic areas, such as guiding principles for safety excellence; consistency between words and actions; cooperation and collaboration; potential inhibitors of mission safety; communications; and employee connection to mission safety. All NASA-specific questions[7] showed improvement compared with the first survey.

FIGURE 10-6. OCDI PERCENTILE[8] SCORES FOR NASA'S GLENN RESEARCH CENTER BEFORE AND SIX MONTHS INTO THE INTERVENTION.

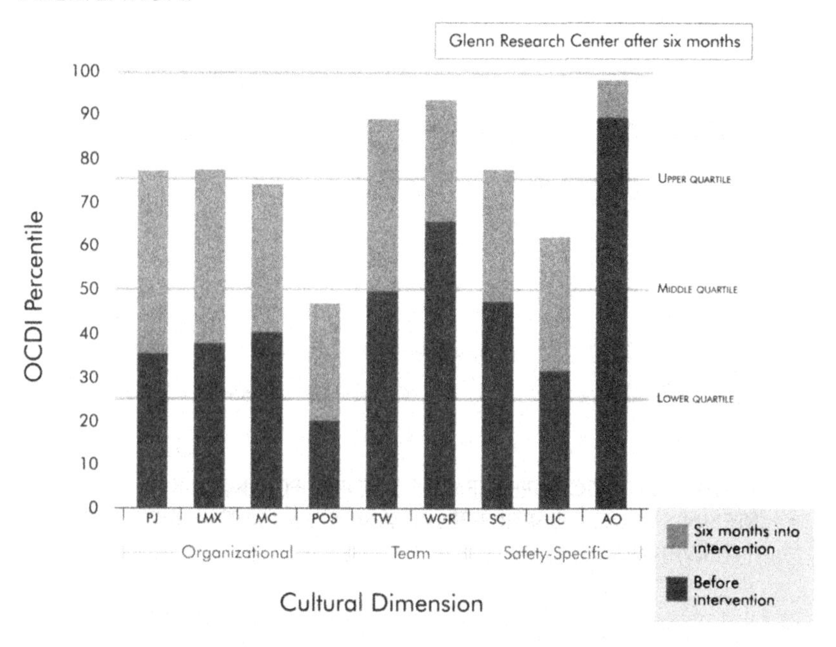

PJ - Procedural Justice · LMX - Leader-Member Exchange · MC - Management Credibility · POS - Perceived Organizational Support
TW - Teamwork · WGR - Workgroup Relations*
SC - Safety Climate · UC - Upward Communication about Safety · AO - Approaching Others about Safety

* WGR (Workgroup Relations) in non-medical fields is the equivalent of TTR (Treatment Team Relations) in healthcare.

7 Only comments mentioning changes to cultural characteristics were counted. Comments about efforts undertaken during the last six months, such as training or meetings, that were descriptors of activities (as opposed to characteristics of culture) were not counted for analysis.

8 Percentile is a rank ordering of the raw score data on a scale of 100 such that the 50th percentile is the middle of the distribution of raw score data.

FIGURE 10-7. OCDI RAW SCORES FOR NASA'S GLENN RESEARCH CENTER BEFORE AND SIX MONTHS INTO THE INTERVENTION.

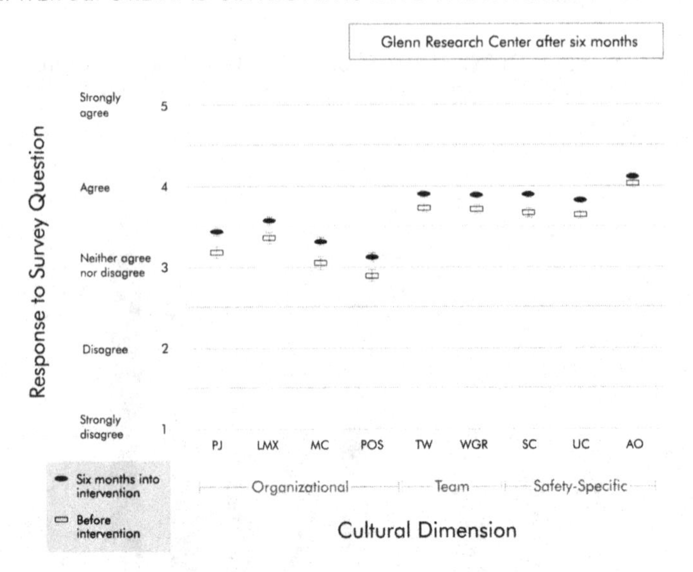

FIGURE 10-8. OCDI PERCENTILE SCORES FOR NASA'S STENNIS SPACE CENTER BEFORE AND SIX MONTHS INTO THE INTERVENTION.

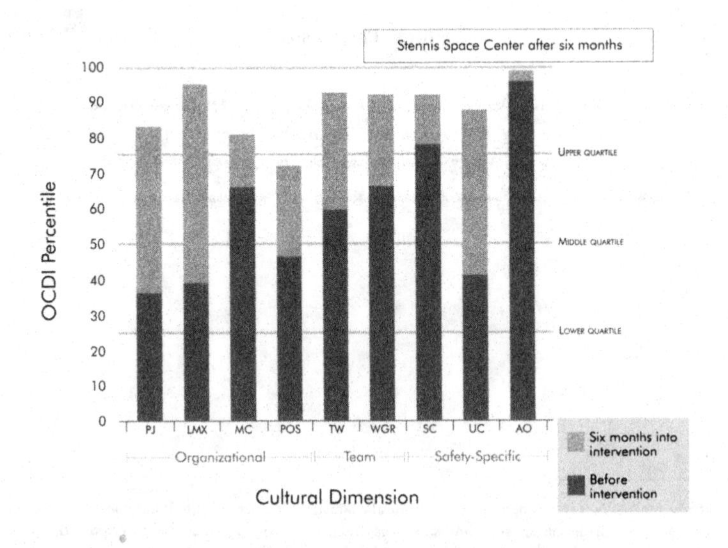

PJ - Procedural Justice · LMX - Leader-Member Exchange · MC - Management Credibility · POS - Perceived Organizational Support

TW - Teamwork · WGR - Workgroup Relations*

SC - Safety Climate · UC - Upward Communication about Safety · AO - Approaching Others about Safety

* WGR (Workgroup Relations) in non-medical fields is the equivalent of TTR (Treatment Team Relations) in healthcare.

FIGURE 10-9. OCDI RAW SCORES FOR NASA'S STENNIS SPACE CENTER BEFORE AND SIX MONTHS INTO THE INTERVENTION.

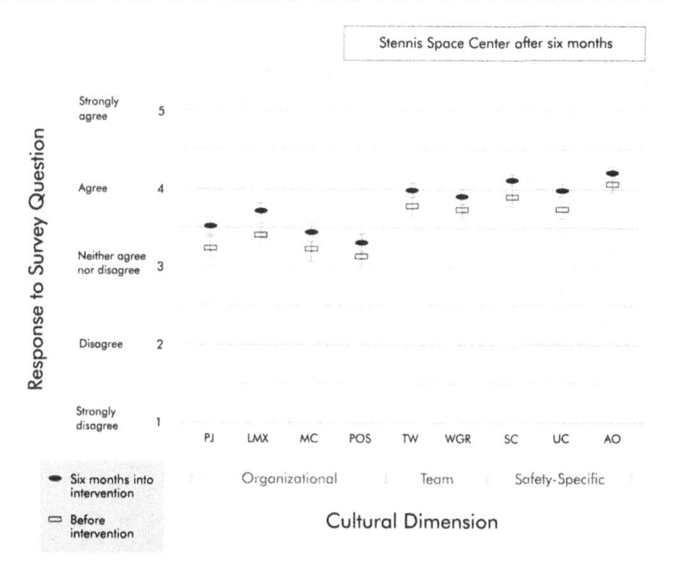

FIGURE 10-10. OCDI PERCENTILE SCORES FOR NASA'S JOHNSON SPACE CENTER BEFORE AND SIX MONTHS INTO THE INTERVENTION.

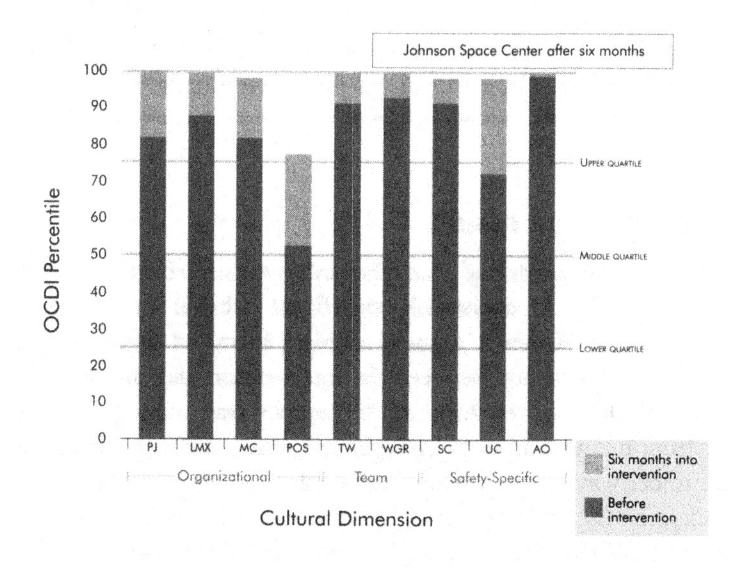

PJ - Procedural Justice · LMX - Leader-Member Exchange · MC - Management Credibility · POS - Perceived Organizational Support
TW - Teamwork · WGR - Workgroup Relations*
SC - Safety Climate · UC - Upward Communication about Safety · AO - Approaching Others about Safety

* WGR (Workgroup Relations) in non-medical fields is the equivalent of TTR (Treatment Team Relations) in healthcare.

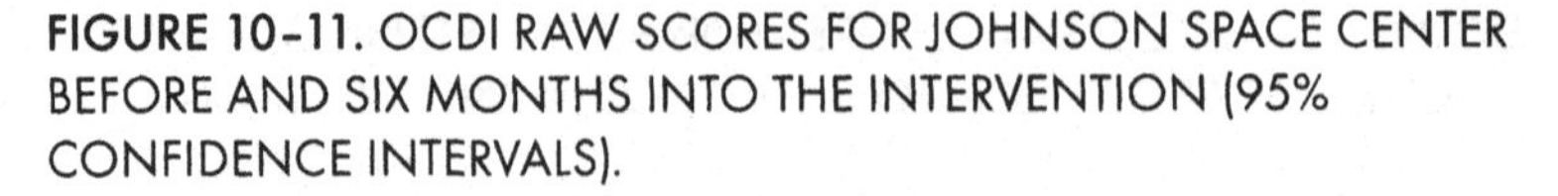

FIGURE 10–11. OCDI RAW SCORES FOR JOHNSON SPACE CENTER BEFORE AND SIX MONTHS INTO THE INTERVENTION (95% CONFIDENCE INTERVALS).

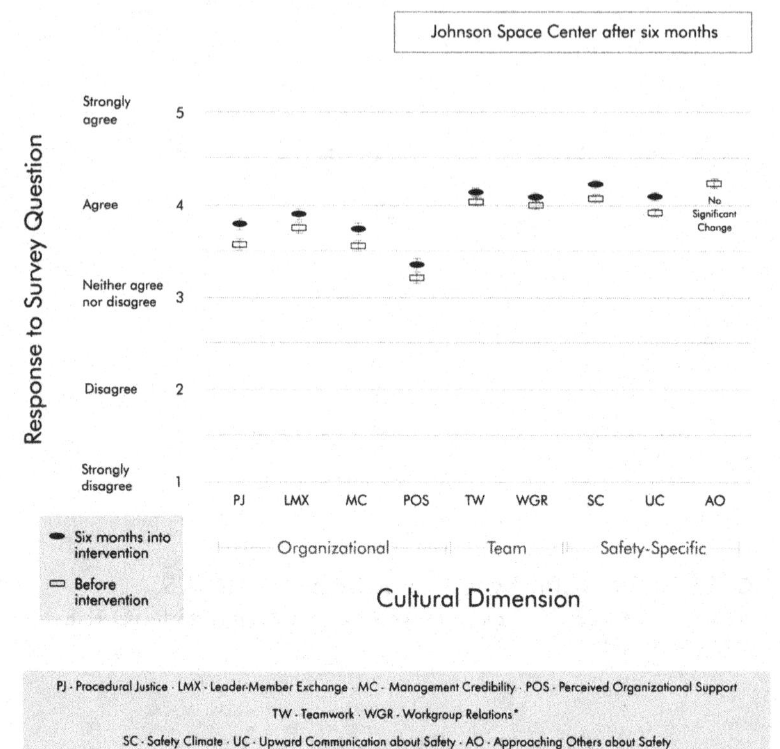

* WGR (Workgroup Relations) in non-medical fields is the equivalent of TTR (Treatment Team Relations) in healthcare.

Interpreting the results

The results demonstrated that the intervention efforts correlate with important and statistically significant cultural improvements in a short time. However, a question might be raised about the specific causal relationship between the intervention and the observed changes in OCDI scores. After all, the agency was under great pressure from many sources to change, and many different initiatives were under way by the time we started the one described here.

Six months into the NASA intervention we collected data at two locations that did not participate in the initial intervention and therefore served as a reference group (Table 10–1). Whereas the centers that participated in the intervention improved significantly, Kennedy and Goddard failed to show such improvement. This suggests that the intervention was chiefly responsible.

TABLE 10-1. OCDI RAW SCORES FROM KENNEDY AND GODDARD SPACE CENTERS, WHICH SERVED AS THE REFERENCE GROUP, SIX MONTHS INTO THE INTERVENTION.

	KENNEDY	GODDARD
Procedural justice (PJ)	3.56	3.51
Leader-member exchange (LMX)	3.70	3.65
Management credibility (MC)	3.44	3.43
Perceived organizational support (POS)	3.27	3.11
Teamwork (TW)	3.93	3.91
Workgroup relations (WGR)	3.90	3.86
Safety climate (SC)	4.03	3.78
Upward communications (UC)	3.94	3.65
Approaching others (AO)	4.14	4.03

NASA's assessment of the BST intervention

Near the end of BST's engagement with NASA an independent firm was asked to evaluate NASA's progress. The firm interviewed NASA personnel from all levels of the organization and produced a lengthy report. Following are excerpts from the executive summary:

> **Opinions of the overall effort**
>
> A majority... 84%... of the NASA personnel interviewed... found the culture change work conducted by BST useful. The effort has helped them to identify problems and has raised the awareness of the importance of open exchanges, effective communication, and leadership. Many have observed a visible change...
>
> Comments from [four individuals who] found the process less useful were not negative toward BST. [Two individuals] said the project was curtailed too soon to be useful; [one said] the schedule imposed by NASA headquarters created resistance that compromised the success of the effort; [one said] hiring BST was a

misinterpretation of the [Columbia Accident Investigation Board] report and NASA's culture did not need to be changed...

There were a variety of opinions as to which aspect of the work was most valuable. Most... participants indicated either the 360-feedback or the behavior observation and feedback were the most valuable...The 360-feedback was valued because it was delivered on an individual basis and did not allow employees to say issues were with others rather than with themselves. The behavioral observation and feedback process provided very fast feedback that reinforced behaviors instead of just giving directions and was well organized, with clear objectives.

Multiple participants indicated that skills training was most valuable to them. They appeared impressed with the documentation that was presented as a basis for the concepts, and the cognitive bias segment appears to have been new material that many found very useful...

Fewer of the NASA personnel interviewed were involved in individual coaching; however, some of those who were [involved in individual coaching] found this to be the most valuable aspect... because they have seen the impact of the effort in their own and other's behavior...

...80%... rated the value of the overall effort above average (a 5, 6, or 7 on a 7-point where 1=poor and 7=excellent)... 72%... rated the benefits their center derived from the work above average. Far fewer... 50%... gave an above average rating to the impact of the overall effort, mostly because of early termination of the project.

Future impact of the effort

...84%... believe the culture change work conducted by BST will result in lasting changes at their center, although some cautioned that the effects could fade without reinforcement or could be negated by a change in leadership at the individual centers. Those who do not believe there will be lasting changes said the work was stopped too quickly to be effective... One [individual] believes leaders cannot be created by any amount of training.

...84%... plan to build on the work done to date. Most said they will continue with refined versions of the components they found most useful...

Reevaluating the effort

...76%... said if they had it to do over, they would engage in this approach again and... 80%... said they would recommend the approach to others. Most often, participants said they would approach the work in a slower fashion, with fewer simultaneous activities. Some would prefer materials to be more tailored for their specific needs.

...92%... would both use BST again and recommend them to others...

Lessons for healthcare

There are many parallels between healthcare and NASA. Like NASA, healthcare has suffered disruptive changes over the last 30 years, stemming in part from resource constraints and shifting priorities. Just as NASA has been said to rely "on past success as a substitute for sound engineering practices,"[9] healthcare continues to do things because "we have always done them that way." Healthcare's culture is authority-based and suffers from both organizational and systemic barriers to upward communication and to airing differences of opinion. And, like NASA, healthcare suffers from organizational fragmentation, competing chains of command, and decision-making processes that frequently operate outside the organization's rules.

Like NASA, healthcare faces complex, daunting safety issues:

- Healthcare involves life and death decisions in the context of highly sophisticated and intricate technology.
- Healthcare has a complex organizational structure that employs far-reaching and intricate management processes and controls.
- The organization of healthcare is tightly hierarchical, operating with a strict top-down authority.

9 Columbia Accident Investigation Board, *Columbia Accident Investigation Board Report* (August 2003), vol. 1, p. 9.

- Healthcare is administered by intelligent, well-educated professionals and enjoys a workforce with the same characteristics.
- Healthcare serves numerous constituencies with diverse, competing agendas.
- In healthcare the motivations of individual employees, contractors, and staff tend to the idealistic and the culture tends to glorify can-do heroes who deliver results no matter the risks.
- Healthcare's success relies heavily on technical expertise and effective communication within and among teams and individuals.

Despite these challenges, healthcare leaders can indeed improve the safety of their organizations by focusing on the lessons learned at NASA, which can be summarized this way: own the problem; believe in fast, positive change; start with a current state analysis; focus on relationships, not tasks; and create the culture.

Own the problem

All too often leaders pay little attention to the culture of their organizations until problems arise. This oversight is unfortunate because culture, if attended to deliberately and proactively, can contribute to the fulfillment of the organization's mission and strategy. A proactive stance in support of safety is a powerful leadership tool for driving the development of a culture that supports safety, and such a culture can itself be a critical driver for continuous hazard reduction.

How often are required safety procedures bypassed in healthcare? How much energy must we expend in an uphill battle to prove that safety is a significant issue? How much does healthcare safety suffer from performance pressure, the equivalent of NASA's long history of having to operate under the motto of "faster, cheaper, better"?[10]

---

10 Philip Zimbardo, *The Lucifer Effect: Understanding How Good People Turn Evil* (New York: Random House, 2007), p. 227. As Zimbardo points out, this politically motivated slogan was one of the factors that led to the first shuttle disaster, that of *Challenger* in 1986.

Positive change can begin quickly

A common refrain among culture change experts is that culture takes a long time to change. There is some truth to this view, but a lot can be accomplished quickly. NASA's ambitious time line had the virtue of relentlessly driving the change effort. Moreover, working against tough deadlines was not foreign to the NASA culture, nor is it foreign to healthcare's culture. The willingness of the NASA leadership to adopt this tough time line communicated their seriousness.

Sean O'Keefe, NASA's 10th administrator, personally sponsored this culture change objective with great interest and passion and powerfully communicated NASA's commitment to change the culture. Personal leadership is required for successful change in organizational culture, and in this case, the highest ranking leader became its champion. In healthcare, many champions have been hard at work at the national level, increasing the visibility of patient safety and making substantial headway. A partial list includes:[11]

- The Institute of Medicine (IOM) brought the crisis in patient safety to the public's attention with its 1999 report, *To Err is Human: Building a Safer Health System,*[12] and later gave us *Crossing the Quality Chasm: A New Health System for the 21st Century.*[13]
- Don Berwick, Lucian Leape, Robert Wachter, Kaveh Shojania, and other physicians have campaigned unceasingly for years to improve patient safety.
- Anesthesiology, an early group to take action for patient safety, and several other medical specialty, pharmacy, and professional associations have created best practice guidelines and other safety measures.
- The U.S. Department of Health and Human Service's Agency for Healthcare Research and Quality (AHRQ)

---

11 With apologies to the many leaders we have not included in this brief list.

12 Linda T. Kohn, Janet M. Corrigan, and Molla S. Donaldson, eds. Committee on Quality of Health Care in America, Institute of Medicine (Washington, DC: National Academy Press).

13 Committee on Quality of Health Care in America, Institute of Medicine (Washington, DC: National Academy Press, 2001).

funded an initial burst of patient safety research after the IOM's 1999 report and continues to host the WebM&M (Morbidity & Mortality Rounds on the Web) conference.

- The National Patient Safety Foundation (NPSF) began working to improve the safety of patients even earlier than the IOM's 1999 report on patient safety. Among its many contributions, the foundation sponsors research in patient safety, a patient safety e-mail discussion forum, and the NPSF Annual Patient Safety Congress.
- The Veterans Healthcare System was an early adopter and innovator in applying lessons from NASA and industry to patient safety.
- The Joint Commission has written patient safety requirements into its accreditation guidelines.
- The Leapfrog Group, an industry consumer consortium, requires that certain quality standards be met before a hospital will be reimbursed for services to employees of the consortium.
- Atul Gawande,[14] Jerome Groopman,[15] and other physician-authors have written powerfully and persuasively about patient safety, educating the public about the complex issues involved and enabling patients to better attend to their own safety.

Start with a current state analysis

Translating this high-level attention and effort to the organizational culture and day-to-day operation of the local hospital and physician's office is a challenge. Here the effort often is less robust and even piecemeal. A more systematic approach starts with a thorough current state analysis.

Assessing the current state is foundational. NASA's current state enjoyed many strengths, and the same is true of healthcare. It is important to clearly understand a culture's strengths because they

---

14 Atul Gawande, *Complications: A Surgeon's Notes on An Imperfect Science* (New York: Metropolitan Books, 2002).

15 Jerome Groopman, *How Doctors Think* (Boston: Houghton Mifflin Co., 2007).

can provide a foundation on which to build, and because uninformed overreliance on existing strengths can create problems.

Adding supplemental questions to the OCDI is a useful and efficient way to obtain additional information about previously identified issues known to be relevant to intervention planning. Interviews and focus groups can also be used to develop the background information needed to interpret OCDI scores fully and to understand them deeply.

NASA's assessment identified significant deficiencies in upward communication. This is also a significant obstacle in healthcare safety. Poor upward communication is often associated with low organizational scores such as NASA's low perceived organizational support. NASA's intervention gives us useful ways to address healthcare's upward communication issues.

Focus on relationships, not tasks

In our experience, highly skilled technical people too often emphasize task orientation to the neglect of relationships—creating an unconscious vulnerability in technical organizations. Unknowingly bringing excessive task orientation into attempted solutions my not only undermine the organization's efforts to correct cultural problems but actually compound them. Demoting safety issues to the status of problem-solving projects is a common example of this dynamic.

In healthcare, efforts to improve safety are frequently undertaken as stand-alone projects without senior sponsorship. This piecemeal approach results in one-off solutions, fragmented implementations, and inconsistent practices across the organization. It overlooks the opportunity to use individual projects as a means of effecting more comprehensive culture change. This mistake tends to befall organizations and professions staffed with smart, highly educated, specialized professionals who spend much of their lives solving problems. A lack of integration in these problem-solving activities results in the neglect or absence of a cultural and organizational umbrella that would enable people to see how the problems all fit together and help them become engaged in their resolution. This situation prevails in many healthcare organizations today.

Finally, highly technical, task-centered organizations can get caught up in paying less attention to the welfare of staff (and patients) than they do to getting the job done, where "the job" is understood in narrow, technical terms. This brings to mind the old saw, "It was a brilliant and masterful surgery. Unfortunately, the patient died."

Create the culture

The leader's role in improving the culture of his or her organization starts with the realization that leaders are already (and always) creating the culture. Healthcare leaders need to assume the responsibility of creating culture deliberately and systematically by identifying and aligning on safety as a strategic priority, developing a leadership safety vision, identifying the gaps between the current state and the vision, developing and implementing a plan to bridge the gaps, and continually renewing the effort based on the analysis of valid data from both upstream and downstream metrics.

All the tools we have provided in this book are based on solid evidence. Their results have been peer-reviewed, and the tools are being extended almost daily in our work with clients in healthcare and other complex industries characterized by rapid technological change. Taken together, the elements of this approach provide a powerful mechanism for producing continuous safety improvement.

BIBLIOGRAPHY

A selection of books, journal articles, and sources for readers who want more in-depth information on the concepts discussed in this book

John Antonakis, Anna T. Cianciolo, and Robert J. Sternberg, eds., *The Nature of Leadership* (Thousand Oaks, CA: Sage Publications, 2004).

Bruce J. Avolio, *Full Leadership Development: Building the Vital Forces in Organizations* (Thousand Oaks, CA: Sage Publications, 1999).

Julian Barling, Catherine Loughlin, and E. Kevin Kelloway, "Development and Test of a Model Linking Safety-Specific Transformational Leadership and Occupational Safety," *Journal of Applied Psychology*, 87 (2002): pp. 488–496.

Bernard M. Bass, "From Transactional to Transformational Leadership: Learning to Share the Vision," *Organizational Dynamics*, 18 (1990): pp. 19–31.

David Bohm, *On Dialogue* (Ojai, CA: David Bohm Seminars, 1990).

BST, *Results Achieved with the Behavioral Accident Prevention Process Technology*, BST Results Studies, 5th ed. (Ojai, CA: BST, 2001). Booklet.

Kerry D. Carson, Paula Phillips Carson, Ram Yallapragada, and C. William Roe, "Teamwork or Interdepartmental Cooperation: Which Is More Important in the Health Care Setting?" *Health Care Manager*, 19 (2001): pp. 39–46.

Jim Collins, *Good to Great: Why Some Companies Make the Leap... and Others Don't* (New York: HarperBusiness, 2001).

Columbia Accident Investigation Board, *Columbia Accident Investigation Board Report*, August 2003, vol. 1.

Corporate Executive Board, *A Compendium of Board Reports: Enhancing Oversight of the Compliance and Ethics Program* (Washington, DC: 2006).

Jacqueline A-M. Coyle-Shapiro and Neil Conway, "Exchange Relationships: Examining Psychological Contracts and Perceived Organizational Support," *Journal of Applied Psychology*, 90 (2005): pp. 774–781.

———·and Ian Kessler and John Purcell, "Exploring Organizationally Directed Citizenship: Reciprocity or 'It's My Job'?" *Journal of Management Studies*, 41 (January 2004): pp. 85–106.

Pat Croskerry, "Achieving Quality in Clinical Decision Making: Cognitive Strategies and Detection of Bias," *Academic Emergency Medicine*, 9 (2002): p. 1184–1204.

Farr A. Curlin, Marshall H. Chin, Sarah A. Sellergren, Chad J. Roach, and John D. Lantos, "The Association of Physicians' Religious Characteristics with Their Attitudes and Self-Reported Behaviors Regarding Religion and Spirituality in the Clinical Encounter," *Medical Care*, 44 (2006): pp. 446–453.

Antonio R. Damasio, *Descartes' Error: Emotion, Reason, and the Human Brain* (New York: G. P. Putnam's Sons, 1994).

Robyn Dawes, *Rational Choice in an Uncertain World* (New York: Harcourt Brace Jovanovich College Publishers, 1988).

Deanne N. Den Hartog, Jaap J. Van Muijen, and Paul L. Koopman, "Transactional versus Transformational Leadership: An Analysis of the MLQ," *Journal of Occupational and Organizational Psychology*, 70 (1997): pp. 19–34.

Sharon S. Dunn and Thomas R. Krause, "Leading to Better Patient Safety, Part 1," *Hospital and Health Networks Online*, May 29, 2007.

———·and Thomas R. Krause, "Leading to Better Patient Safety, Part 2," *Hospital and Health Networks Online*, June 5, 2007.

Judith A. Erickson, "The Relationship between Corporate Culture and Safety Performance," *Professional Safety*, 42 (May 1997): pp. 29–33.

Jonathan Evans, *Bias in Human Reasoning: Causes and Consequences*, Essays in Cognitive Psychology (London: Psychology Press, 1990).

Gail T. Fairhurst, L. Edna Rogers, and Robert A. Sarr, "Manager-Subordinate Control Patterns and Judgments about the Relationship," in Margaret L. McLaughlin, ed., *Communication Yearbook 10* (Beverly Hills, CA: Sage Publications, 1987), pp. 395–415.

Gerald R. Ferris, "Role of Leadership in the Employee Withdrawal Process: A Constructive Replication," *Journal of Applied Psychology*, 70 (1985), pp. 777–781.

Howard Gardner, *Five Minds for the Future* (Boston: Harvard Business School Press, 2006).

Atul Gawande, *Complications: A Surgeon's Notes on an Imperfect Science* (New York: Picador, 2003).

Jerald Greenberg and Russell Cropanzano, eds., *Advances in Organizational Justice* (Stanford, CA: Stanford University Press, 2001).

Jerome Groopman, *How Doctors Think* (Boston: Houghton Mifflin, 2007).

John S. Hammond, Ralph L. Keeney, and Howard Raiffa, "The Hidden Traps in Decision Making," *Harvard Business Review*, 76 (1998): pp. 47–58.

Marc D. Hauser, *Moral Minds: How Nature Designed Our Universal Sense of Right and Wrong* (New York: HarperCollins, 2006).

H. W. Heinrich, *Industrial Accident Prevention: A Scientific Approach*, 4th ed. (New York: McGraw Hill, 1959).

John H. Hidley, "Critical Success Factors for Behavior-Based Safety," *Professional Safety*, July 1998: pp. 30–34.

C. K. Hofling et al., "An Experimental Study in Nurse-Physician Relationships," *Journal of Nervous and Mental Disease*, 143 (1966): pp. 171–180.

David A. Hofmann and F. P. Morgeson, "Safety-Related Behavior as a Social Exchange: The Role of Perceived Organizational Support and Leader-Member Exchange," *Journal of Applied Psychology*, 84 (1999): p. 286–296.

David A. Hofmann and Adam Stetzer, "The Role of Safety Climate and Communication in Accident Interpretation: Implications for Learning from Negative Events," *Academy of Management Journal*, 41 (December 1998): pp. 644–657.

Robert Hogan, Gordon J. Curphy, and Joyce Hogan, "What We Know about Leadership: Effectiveness and Personality," *American Psychologist*, 49 (1994): pp. 493–504.

Robin Hogarth, *Judgment and Choice* (New York: John Wiley & Sons, 1980).

Jim Huggett and Thomas R. Krause, "Executive Coaching for the Safety Leader" (Ojai, CA: BST, 2005). White paper.

G. M. Hurtz and J. J. Donovan, "Personality and Job Performance: The Big Five Revisited," *Journal of Applied Psychology*, 85 (2002): pp. 869–879.

Mark A. Huselid, "The Impact of Human Resource Management Practices on Turnover, Productivity, and Corporate Financial Performance," *Academy of Management Journal*, 38 (1995): pp. 635–672.

Institute of Medicine, *To Err Is Human: Building a Safer Health System*, eds. L. T. Kohn et al. (Washington, DC: National Academy Press, 1999).

Joint Commission, "Root Causes of Sentinel Events," Viewable at www.jointcommission.org.

T. A. Judge, J. E. Bono, R. Ilies, and M. W. Gerhardt, "Personality and Leadership: A Qualitative and Quantitative Review," *Journal of Applied Psychology*, 87 (2002): pp. 765–780.

T. A. Judge, C. A. Higgens, C. J. Thoreson, and M. R. Barrick, "The Big Five Personality Traits, General Mental Ability, and Career Success across the Life Span," *Personnel Psychology*, 52 (1999): pp. 621–652.

Daniel Kahneman, Paul Slovic, and Amos Tversky, eds., *Judgment under Uncertainty: Heuristics and Biases* (Cambridge: Cambridge University Press, 1982).

Tal Katz-Navon, Eitan Naveh, and Zvi Stern, "Safety Climate in Healthcare Organizations: A Multidimensional Approach," *Academy of Management Journal*, 48 (2005): pp. 1075–1089.

Alan E. Kazdin, *Behavioral Modification in Applied Settings*, 6th ed. (Belmont, CA: Wadsworth Publishing, 2000).

Jill R. Kickul, Lisa K. Gundry, and Margaret Posig, "Does Trust Matter? The Relationship between Equity Sensitivity and Perceived Organizational Justice," *Journal of Business Ethics*, 56 (February 2005), pp. 205–218.

Mary A. Konovsky and S. Douglas Pugh, "Citizenship Behavior and Social Exchange," *Academy of Management Journal*, 37 (June 1994): pp. 656–669.

John P. Kotter, *Leading Change* (Boston: Harvard Business School Press, 1996).

James M. Kouzes and Barry Z. Posner, *The Leadership Challenge* (San Francisco: Jossey-Bass, 1995).

Daniel J. Koys, "The Effects of Employee Satisfaction, Organizational Citizenship Behavior, and Turnover on Organizational Effectiveness: A Unit-Level, Longitudinal Study," *Personnel Psychology*, 54 (March 2001): pp. 101–114.

Annamarie Krackow and Thomas Blass, "When Nurses Obey or Defy Inappropriate Physician Orders: Attributional Differences," *Journal of Social Behavior and Personality*, 10 (1995): pp. 585–595.

Gerald A. Kraines, *Accountability Leadership: How to Strengthen Productivity through Sound Managerial Leadership* (Franklin Lakes, NJ: Career Press, 2001).

Thomas R. Krause, *Employee-Driven Systems for Safe Behavior: Integrating Behavioral and Statistical Methodologies* (New York: Van Nostrand Reinhold, 1995).

———. *Leading with Safety* (Hoboken, NJ: John Wiley & Sons, 2005).

———. *The Behavior-Based Safety Process: Managing Involvement for an Injury-Free Culture*, 2nd ed. (New York: John Wiley & Sons, 1997).

———. and John Balkcom, "Creating a Culture of Responsibility," *Directors and Boards*, 31 (December 2006): pp. 36–39.

———. and Sharon Dunn, "Twelve Questions Every CEO Should Answer about Patient Safety," *Patient Safety & Quality Healthcare*, November/December 2006. Supplement.

———. and Sharon Dunn, "How Are Leadership and Culture Related to Patient Safety?" (Ojai, CA: BST, 2006). White paper.

———·and K. J. Seymour and K. C. M. Sloat, "Long-Term Evaluation of a Behavior-Based Method for Improving Safety Performance: A Meta-Analysis of 73 Interrupted Time-Series Replications," *Safety Science*, 32 (1999): pp. 1–18.

———·and Thomas Weekley, "Safety Leadership: A Four-Factor Model for Establishing a High-Functioning Organization," *Professional Safety*, 50 (November 2005): pp. 34–40.

D. Kriebel, "Occupational Injuries: Factors Associated with Frequency and Severity," *International Archive of Occupational and Environmental Health*, 50 (1982): pp. 209–218.

Fred A. Manuele, *Heinrich Revisited: Truisms or Myths* (Itasca: National Safety Council Press, 2002).

———·"Injury Ratios: An Alternative Approach for Safety Professionals," *Professional Safety*, 49 (February 2004): pp. 22–30.

Barbara Mark and David A. Hofmann, "An Investigation of the Relationship between Safety Climate and Medication Errors as Well as Other Nurse and Patient Outcomes," *Personnel Psychology*, 59 (2006): pp. 847–869.

Douglas McCarthy and David Blumenthal, *Committed to Safety: Ten Case Studies on Reducing Harm to Patients* (New York: The Commonwealth Fund, April 2006).

Robert H. Moorman, Gerald L. Blakely, and Brian P. Niehoff, "Does Perceived Organizational Support Mediate the Relationship between Procedural Justice and Organizational Citizenship Behavior?" *Academy of Management Journal*, 41 (1998): pp. 351–357.

Michael K. Mount, Murry R. Barrick, and Greg L. Stewart, "Five-Factor Model of Personality and Performance in Jobs Involving Interpersonal Interaction," *Human Performance*, 11 (1998): pp. 145–165.

Richard E. Nisbett, *The Geography of Thought: How Asians and Westerners Think Differently... and Why* (New York: Free Press, 2003).

———·and Lee Ross, *Human Inference: Strategies and Shortcomings of Human Judgment* (Englewood Cliffs, NJ: Prentice-Hall, 1980).

Dan Petersen, *Safety Management: A Human Approach*, 2nd ed. (Des Plaines, IL: American Society of Safety Engineers, 1998).

———."The Four Cs of Safety: Culture, Competency, Consequences, and Continuous Improvement," *Professional Safety* (April 1998), pp. 32–34.

Steven Pinker, *The Language Instinct: How the Mind Creates Language* (New York: HarperCollins, 2004).

Michael A. Roberto, "Lessons from Everest: The Interaction of Cognitive Bias, Psychological Safety, and System Complexity," *California Management Review,* 45 (Fall 2002): pp. 136–158.

Cynthia P. Ruppel and Susan J. Harrington, "The Relationship of Communication, Ethical Work Climate, and Trust to Commitment and Innovation," *Journal of Business Ethics,* 25 (June 2000): pp. 313–328.

Kaveh G. Shojania et al., eds., *Making Health Care Safer: A Critical Analysis of Patient Safety Practices* (Rockville, MD: Agency for Healthcare Research and Quality, 2001). AHRQ Publication No. 01-E058. Available at www.ahrq.gov.

R. Scott Stricoff and Kristen J. Seymour, "Applying Six Sigma to Safety" (Ojai, CA: BST, 2002). White paper.

Beth Sulzer-Azaroff and Roy G. Mayer, *Behavior Analysis for Lasting Change,* 2nd ed. (Belmont, CA: Wadsworth Publishing, 1991).

Judith W. Tansky and Debra J. Cohen, "The Relationship between Organizational Support, Employee Development, and Organizational Commitment: An Empirical Study," *Human Resource Development Quarterly,* 12 (2001): pp. 285–300.

Benjamin B. Tregoe and John W. Zimmerman, *Top Management Strategy: What It Is and How to Make It Work* (New York: Simon and Schuster, 1980).

Jatin G. Vaidya, Elizabeth K. Gray, Jeffrey R. Haig, and David Watson, "On the Temporal Stability of Personality: Evidence for Differential Stability and the Role of Life Experiences," *Journal of Personal and Social Psychology,* 83 (December 2002): pp. 1469–1484.

Robert M. Wachter and Kaveh G. Shojania, *Internal Bleeding: The Truth Behind America's Epidemic of Medical Mistakes* (New York: Rugged Land, 2005).

Sandy J. Wayne, Robert C. Liden, Maria L. Kraimer, and Isabel K. Graf, "The Role of Human Capital, Motivation, and Supervisor Sponsorship in Predicting Career Success," *Journal of Organizational Behavior,* 20 (1999): pp. 577–595.

Steve Williams, "The Effects of Distributive and Procedural Justice on Performance," *The Journal of Psychology,* 133 (1999): pp. 183–193.

Philip Zimbardo, *The Lucifer Effect: Understanding How Good People Turn Evil* (New York: Random House, 2007).

Dov Zohar, "A Group-Level Model of Safety Climate: Testing the Effect of Group Climate on Microaccidents in Manufacturing Jobs," *Journal of Applied Psychology,* 85 (2000): pp. 587–596.

———."Promoting the Use of Personal Protective Equipment by Behavior Modification Techniques," *Journal of Safety Research,* 12 (1980): pp. 78–85.

———."The Effects of Leadership Dimensions, Safety Climate, and Assigned Priorities on Minor Injuries in Work Groups," *Journal of Organizational Behavior,* 23 (February 2002): p. 75–92.

———.and Nahum Nussfeld, "Modifying Earplug Wearing Behavior by Behaviour Modification Techniques: An Empirical Evaluation," *Journal of Organizational Behaviour Management,* 3 (1981): pp. 41–52.

INDEX

360-degree diagnostic instrument, 95, 118
360-degree feedback, 7, 199

ABC analysis, 141–145, 146
  step 1, 141
  step 2, 144
  step 3, 145
Accountability, 124, 131
  exercise, 132
Acknowledgment, 130
Action, 123
Action orientation, 127
  exercise, 128
Action plan, creating, 145
Actor-observer bias, 162, 164
Administrator
  challenges of, 27
  responsibilities of, 40, 43, 51
Adverse event, 36, 150, 158, 160, 164
  analysis of, 55
  cost of, 15
  defined, 21
  preventability of, 21, 23
  relationship between serious and less serious, 38
  relationship to exposure, 36–39
  signaling overlooked exposures, 41
Agency for Healthcare Research and Quality, 13, 66, 267
Alcoa, 2, 190
  employee fatalities, 3
  workplace safety, 2
Analyzing
  current state, 227
  desired behaviors, 141, 144
  undesired behaviors, 141–143
Anchoring bias, 156, 158, 170
Antecedent, 139
Antecedent-behavior-consequence analysis—*see* ABC analysis
Antecedents, behaviors, and consequences, 139
Applied behavior analysis, 136, 146
  applying principles, 146
  principles, 139
Approaching others, 70, 82, 174
  at NASA, 245
  defined, 65
Assessing groups, 95
Attribution errors, 163, 165
Availability/nonavailability bias, 156, 158

Behavior, 7
  compensating for, 107
  defined, 139
  desired and undesired, 141
Behavior change, 136
  defined, 137
Behavior measurement process, 223
Behavior observation and feedback, 222, 224
  at NASA, 256
Benchmarking, 191
Berwick, Don, 267

Best practices
in Safety Leadership Model, 94
protocols, 25
safety leadership, 120
Bias identification, exercise, 171
Biases
of data selection, 155
of data use, 162–163
Big Five, 99–101, 107
and career success, 102
and leadership effectiveness, 102
and leadership emergence, 102
insights for safety improvement, 103–105
measurement of, 102
Blame, 166
avoidance of, 22, 52, 57, 215
Blame-free exploration, 174
Blood transfusion, case study, 171
Blueprint for Healthcare Safety Excellence, 32, 55, 60, 136, 197, 198, 201, 205, 227
Body language, 180
Business considerations, 15

Captain of the ship
attitude, 23
doctrine, 22
Case studies
blood transfusion, 171
cognitive bias
in manufacturing, 169–170
Commonwealth Fund, 18–19
Exemplar HealthNet, 225–233, 233–237
glutaraldehyde error, 22, 49, 55–58
manufacturing plant, 175
patient with leukemia, 11–12, 20
Change
efforts, 85
facilitating, 86
resistance to, 86
strategy at NASA, 252
Changing behavior, 136
exercise, 134
with applied behavior analysis, 136
Channels of influence, 69
Character building, 123
Charter, 207
Chief safety officer, 200
Chisel, as metaphor for healthcare delivery system, 54–55, 57
Climate—*see* safety climate
Clinical protocols, 25
Coach, 141
Coalition, creation of, 5
Cognitive bias, 150, 198
and healthcare safety, 153
defined, 150
detecting, 175
effect on culture, 169
effect on leadership behavior, 169
learning from mistakes, 174
literature on, 151
manufacturing case study, 169
self-monitoring for, 172
unconscious quality of, 154
Collaboration, 124, 128
exercise, 129
Collegiality, 101, 105
Collins, Jim, 192
Columbia Accident Investigation Board, 1, 240, 243, 244, 264
Columbia—*see* space shuttle
Commission bias, 162, 163
Commonwealth Fund, 17–19
Communication, 124, 129
at NASA, 257
errors, 167
exercise, 129
Compassion, 107, 108
Confederation, healthcare organization as, 2, 4
Confirmation bias, 156, 159, 171
Conscientiousness, 101, 105
Consequence, 139
defined, 139
financial, 147
likelihood, 140
positive and negative, 146
potency, 140
significance, 140
soon, certain, positive, 140
timing of, 140
Continuous safety improvement, 211, 270

Core ideology, 192
Corporate Executive Board, 185
Credibility, 123, 126
 exercise, 127
Cultural data, gathering, 201
Cultural dimension
 and focus, 70
 and safety outcomes, 70, 71
 identification exercise, 83
Culture, 6. *Also see* organizational culture
 and the working interface, 49
 as a consequence of leadership, 6, 90
 assessment, at NASA, 245
 correlation with leadership best practices, 97
 defined, 15
 four pillars of, 72
 measurement of, 15, 63
 pace of change, 52
 predicting safety outcomes, 6
 versus climate, 48
Culture change, 164, 182, 211, 223
 and leadership behavior, 52
 launching, 210
 plan, at NASA, 251
 process, at NASA, 254
Culture of blame, avoiding, 186
Current state analysis, 200, 268

Data selection, 172
 biases, 155
Data use biases, 162–163
Decision making, protecting from bias, 150
Déformation professionnelle bias, 162, 164, 171, 172
Dehumanization, 162, 168
Delegation, 206
 of leadership and culture building, 51
 of systems issues, 51
Desired behaviors, 141
Desired future state, 125
Detached concern, 162, 168
Diagnostic instrument, versus informational survey, 65, 66
Dialogue, 212
 defined, 196
 encouragement of, 174
Distinguished Hospitals for Patient Safety, 17
DuPont, 190

Emotional resilience, 99, 104
Employee perceptions, 179
Employee safety
 benefits of, 5
 versus patient safety, 111
Employee satisfaction, as outcome of improved safety, 5
Envisioned future, 192
Epistemic arrogance, 163, 166
Errors of judgment, 150
Establishing goals, 190
Ethical
 commitment, 51
 considerations, 14
 errors, 7
 leadership, 195, 196, 197
Ethics, 194, 195
 as motivation, 14
Exercises
 Accountability, 132
 Action orientation, 128
 Changing behavior, 134
 Collaboration, 129
 Communication, 129
 Credibility, 127
 Freedom from bias, 175
 Identify the biases, 171
 Identify the cultural dimension, 83
 Recognition and feedback, 131
 Vision, 125
Exposure to hazard, 34, 37, 39, 160, 186, 224
 controlling, 43
 defined, 36
 reducing, 179
 relationship to adverse events, 36–37
 signaling need for action, 41
Extroversion, 100, 104
ExxonMobil, 2
 employee fatalities, 3
 workplace safety, 2

Fact/value confusion, 156, 161, 173
False consensus effect, 162, 167

Feedback, 222
Follow-up, importance of, 47
Four pillars of culture, 72
Framing the issue, 172
Free agents, 40
Freedom from bias, exercise, 175
Future state, at NASA, 251

Gaining alignment, 225
Gathering data, 203
  cultural, 228
  hazard, 230
  leadership, 227
Gawande, Atul, 268
Goals, establishing, 190
Groopman, Jerome, 167, 268
Group assessments, 95
Group attribution error, 162, 164, 165

Hazard data, gathering, 203
Hazard identification, 223
Healthcare organizations
  as loose confederations, 2
  fragmentation of, 15
  pliability of, 8
Healthcare safety,
  and cognitive bias, 153
Healthcare safety–enabling elements, 32, 121, 205, 216
  analysis of in adverse event, 56
  auditing, 44
  defined, 34, 44
Healthcare safety leading indicators, 41–42
Healthcare system, 12, 13
  lack of, 29
HealthGrades, 17
Heparin overdose, 40
Hindsight bias, 162, 166, 171
Hospital-acquired infections, 28, 50, 187, 188
Human element, as weak link, 14
Human error, 161
Hurricane Katrina, 218

ICU checklist, 190
Illusion of transparency, 162, 167
Implementation strategy, at NASA, 254
Incentives, material, 147
Incidents, 185
In-group bias, 162, 167
Initiative, communication of, 206
Injury-free culture, 39
Injury rates, and OCDI scores, 66, 67
Institute of Medicine, 4, 13, 21, 267
Interpreting results, at NASA, 262
Intervention
  aligning systems, 237
  at NASA, 250
  behavior observation and feedback, 237
  by group, 233
  design, 178
  executive coaching, 232
  impact, at NASA, 264
  leadership assessment, 232
  Leading with Safety workshops, 236
  NASA assessment of, 263
  periodic cultural reevaluation, 237
  physician engagement, 232
  plan, 205
  plan development, 231
  targeted training, 236
  training in systems thinking, 236
Interview findings, at NASA, 247
Intrinsic value of others, 110, 123

Job satisfaction, 76
Joint Commission, 13, 268

Lagging indicators, defined, 41
Lake Wobegon effect, 163, 164
LDI, 7, 117, 118, 133, 137, 174, 203, 227
  360-degree, 95
  and best practices, 133
  as assessment tool, 97
  measuring leadership best practices with, 133
  measuring leadership with, 95
  reliability of, 95
  validation of, 95
Leader
  as culture creator, 51, 179
  emotional commitment, 107

employee perceptions of, 133
personality, 98
Leader-member exchange, 70, 72, 174
at NASA, 245
defined, 64, 74
Leadership, 4
alignment of, 4, 15, 52, 183, 184
analysis of in adverse event, 56
data gathering, 203
defined, 34
effect on culture, 90
measurement of with LDI, 95
of culture change, 167
overreaction, 41
qualities, 90
relationship to culture, 15
relationship to safety, 15
style, 112
subordinate trust in, 74
values, 98, 107
versus management, 51, 121–122
Leadership behavior, 112
relationship to safety outcomes, 7
shaping culture, 92
Leadership best practices, 120, 174. *also see* vision, credibility, action orientation, collaboration, communication, recognition and feedback, accountability
correlation with culture, 97
correlation with transformational leadership, 96
listed, 122
versus leadership style, 96
Leadership coaching, 232, 234–235
at NASA, 255
Leadership credibility, 70, 72, 174
defined, 64, 74
Leadership Diagnostic Instrument—see LDI
Leadership practices, at NASA, 243
Leadership style, 94, 112
cultivating, 118
in Safety Leadership Model, 94
versus leadership best practices, 96
Leading indicators, 189
board review of, 46
defined, 41
systems for, 46
versus lagging indicators, 41
Leading with Safety process, 180–181
delegation of, 180, 210
phase I, 181
phase II, 181, 210
step 1, 183
step 2, 191
step 3, 200
step 4, 204
step 5, 212
step 6, 216
step 7, 222
step 8, 224
Leape, Lucian, 267
Leapfrog Group, 268
Learning orientation, 100, 104

Malpractice, 53
effect on physician morale, 24
Management
versus leadership, 121–122
Managers
engagement of, 215
safety role of, 215
Manufacturing, cognitive bias case study, 169
Mayo Clinic, 212
Measuring
culture, 63
leadership, 95
leadership best practices with LDI, 133
performance, 138
Mechanism, 91, 136, 146, 150, 217, 219, 224
defined, 35
Medical errors, 168
physician role in, 23
prevention of, 38
Medication dispensing machine overrides of, 6, 7
Momentum bias, 163, 164
Mount Everest tragedy, 151, 160
Multirater feedback, at NASA, 256
NASA, 1
after Columbia, 240–265
approach to transformation, 243

NASA (*continued*)
assessment conclusions, 249
behavior observation and feedback, 256
change strategy, 252
communication, 257
core values, 249
culture and climate assessment, 245
culture change plan, 251
culture change process, 254
impediments to upward communication, 247
implementation strategy, 254
interpreting the results, 262
intervention, 250
intervention assessment, 263
intervention impact, 264
interview findings, 247
leadership coaching, 255
lessons for healthcare, 265–270
multirater feedback, 256
OCDI findings, 245–247
OCDI results, 259–263
results, 257
skills training, 256
supervisor behaviors, 253
vision of future state, 251
National Aeronautics and Space Administration—*see* NASA
National Patient Safety Foundation, 13, 46, 268
Near-miss reporting, 42
Negative bias, 156, 159
Negativity/positivity effect, 163, 167
Nonverbal responses, 179
Nurses, 3, 6, 7
and medical error reduction, 26
and physician orders, 23
morale, 26, 27
overtime, 175
shortage of, 27

OCDI, 63, 157, 198, 201, 228
database, 68
dimensions and safety outcomes, 68
findings at NASA, 245–247
organizational dimensions, 71
results at NASA, 259–263
safety-specific dimensions, 71
scores and injury rates, 66, 67
scores and safety performance, 86
team dimensions, 71
O'Keefe, Sean, 240, 243, 267
O'Neill, Paul, 190
One-off approach, limits of, 18, 32
Optimism bias, 163, 166
Organizational Culture Diagnostic Instrument—*see* OCDI
Organizational culture, 2, 14, 30, 121
alignment of, 15
analysis of in adverse event, 56
at NASA, 240
attributes of, 62
defined, 34, 48
health indicators, 21
in Safety Leadership Model, 94
measurability of, 48
nine dimensions of, 62, 64
relationship to safety and leadership, 1
speed of change, 49
versus safety climate, 34
Organizational dimensions, 70, 71, 201
and favorable/unfavorable perceptions, 77
as predictor of safety outcomes, 72
defined, 72
high OCDI scores , 77
low OCDI scores, 77
Organizational safety, 1
Organizational sustaining systems, 121, 205, 218
analysis of in adverse event, 56
defined, 34, 45
in Leading with Safety process, 219
Outcome measures, 160, 205
Overconfidence bias, 153, 163, 166, 171, 176
Owning the problem, 266

Patient safety
benefits of, 5, 19
common interest in, 15
determining, 9–30

director, 220
indicators, 17
interventions, 18–19
organizational causes, 34
retreat, 182
versus employee safety, 5, 111
vision, defined, 191
workshops, 211
Patient Safety Academy, 181, 182, 201, 208, 236, 237
attendees, 183
time frame, 182
Patient Safety Discussion Forum, 46
Patient satisfaction, as outcome of improved safety, 5
Perceived organizational support, 70, 72, 75, 157, 174
at NASA, 246
defined, 64
Performance
management systems, 46
metrics, 188
variability, 138
Personal safety ethic, 93, 98
in Safety Leadership Model, 93
Personal values, 108
Personality
assessment instruments, 102
attributes, 99–101
relevance to safety leadership, 99–101
research, 108
Physician
as free agent, 27
as safety leader, 23
decision making, 25
nurse complaints about, 27
Positivity/negativity effect, 163, 167
Preventable adverse event
defined, 36
Prevention, 189
Proactive approach, 189
Problem solving, 222
Procedural justice, 70, 72, 157, 174
characteristics of, 73
defined, 64
Profitability, 107
Projection bias, 163, 167, 172
Pronovost, Peter, 190
Prototyping, 154, 156
Reason, James, 54
Reason's chisel, 54–55, 57
Recency bias, 153, 156, 158, 170, 173
Recognition and feedback, 124, 130
exercise, 131
Reinforcement, 140
Reinforcer, 140
Relationship focus, 269
Resistance to change, 10, 16, 40, 86, 166
Resource allocation, 206
Retreating from zebras bias, 156, 159, 170
Rewards and recognition, 220
Root cause analysis, 6, 44, 150, 159, 218
as safety-enabling element, 34
use of findings, 47
Rosy retrospection, 163, 166

Safety
as core value, 185
as ethical issue, 185
as good strategy, 16
as personal value, 108
as strategic priority, 183, 184
audit, 169
best practices, 42
champions, 213
dimensions, 201
function, 220
gap, 137
incentive schemes, 130
metrics, 218
pyramid, 36
regulations, 16
Safety climate, 2, 29, 30, 49, 54, 70, 80, 114, 121
at NASA, 240, 244
defined, 15, 49, 65
measurement of, 15
predicting patient and nurse safety, 68
predicting patient and nurse satisfaction, 68
speed of change, 49
versus organizational culture, 34, 48

Safety-critical
behaviors, 136, 146, 187, 220
practices, 137
Safety culture—as misnomer, 1
Safety data, 217
collection of, 187
Safety-enabling elements—*see* healthcare safety–enabling elements
Safety improvement
achieving, 72
intervention design, 178
motives, 107
objective, 51
Safety leader
motivation, 10, 14
qualities of, 90
responsibilities of, 51
role of, 10, 30
Safety Leadership Model, 93, 107, 111, 120, 134
reliability of, 95
rings in, 92
validation of, 95
Safety leadership skills
acquiring, 58
Safety outcomes
and cultural dimensions, 70
organizational influence on, 87
relationship to leadership, 1, 6, 7
relationship to organizational culture, 1, 6, 7
Safety roadblocks, 16
administrator point of view, 27
nurse point of view, 26
patient point of view, 19
physician point of view, 21
Safety-specific dimensions, 70, 71, 80
and favorable/unfavorable perceptions, 83
high OCDI scores, 83
low OCDI scores, 83
Safety systems, 46
and preventable adverse events, 47
delegation of, 46
Safety team charter, 207
Safety vision
and ethics, 194
employee benefits from, 194
rewards of, 193
sharing of, 197, 198, 199
Safety-enabling elements—*see* healthcare safety–enabling elements
Sample bias, 156, 160
Sarbanes-Oxley Act, 185
Search satisficing bias, 156, 161
Seat belt use, 39, 41
Selective perception, 156, 159, 173
Self-serving bias, 163, 164
Sentinel events, 160, 186, 189, 225
relationship between frequency and severity, 37
Shojania, Kaveh, 13, 267
Silent evidence bias, 156, 159, 170
Six Sigma, 47
Skills training, at NASA, 256
Social exchange theory, 75, 76
Soon, certain, and positive
consequences, 140
feedback, 130
Southwest Airlines, 184
Space shuttle *Columbia*, 1, 240
accident cause, 1
Staff engagement, 214
Statistical variability, 41
Status quo bias, 156, 161, 170, 173, 175
Strategic objectives
and safety, 188
Strategy, 6
defined, 183
Sunk cost bias, 153, 156, 160, 175
Supervisor behaviors
at NASA, 253
Supervisors, engagement of, 214
Surveys
limits of, 65
versus diagnostic instruments, 65, 66
Sustaining mechanisms, 224
Systems
creating, 58
focus, limits of, 48
issues, delegation of, 51
performance, 5
realignment, 216
thinking, 5, 23

Team data, gathering, 204
Team dimensions, 71, 78, 201
 and favorable/unfavorable perceptions, 79
 high OCDI scores, 79
 low OCDI scores, 79
Team relations, 6
Teamwork, 70, 78
 at NASA, 245
 defined, 64
Third-party payers, 13, 26
Time frame
 phase I, 182
 phase II, 211
Tone at the top, 185, 186
Total Quality Management, 47
Tragedy on Mount Everest, 151
Transactional leadership, 94, 112–113
 versus transformational leadership, 115–116
Transformational leadership, 74, 94, 113–118, 174
 and culture change, 117
 and safety performance, 114
 characteristics of, 117
 correlation with leadership best practices, 96
 skills, 117–118
 versus transactional leadership, 115–116
Transparency, 129
 organizational, 212
 to patients, 213
Treatment team, 116
 perceptions, 179
 relationships, 69
 relationships with superiors, 72
Treatment team relations, 70, 79
 defined, 64
Trust, development of, 74
Trustworthiness, 75

Undesired behaviors, 141
Upstream and downstream measures, 187
Upward communication, 70, 82, 157, 174
 at NASA, 246, 247
 defined, 65

Values, 107
 extrinsic, 108
 imposition of, 108
 intrinsic, 108, 110, 114
 shared, 108
Valuing safety, 111
Veterans Healthcare System, 268
Vice president of safety, 200, 220
Virtues, 120
Vision, 123, 124
 development of, 226
 exercise, 125

Wachter, Robert, 13, 267
Wishful thinking bias, 163, 167, 170, 172
Workgroup relations
 at NASA, 245
Working interface, 121, 224
 and culture, 49
 assessment of, 42
 defined, 34, 35, 43
 in healthcare, 43
 venues in, 43
Wrong-site surgery, 45

Zero incidents, as goal, 3
Zimbardo, Philip, 165, 174, 195

Printed in the United States
By Bookmasters